WIND FORCES

Wind Forces

National Council of Architectural Registration Boards
1735 New York Avenue, N.W., Suite 700
Washington, DC 20006
(202) 783-6500
http://www.ncarb.org

ISBN 0-941575-29-2
Printed in the United States of America

This monograph was published in July 1998.

ABOUT THE AUTHORS

Del Ward and Stan Crawley have been associates and friends throughout their careers. Both are architects (Crawley also is a structural engineer), and both are in architectural education. They were principals and partners in an architecture and engineering firm in Salt Lake City, UT, and they have collaborated several times as co-authors of technical material for architects. *Seismic and Wind Loads in Architectural Design* (1990), a reference handbook, printed and distributed by the American Institute of Architects, preceded this monograph.

Ward came to the University of Utah in 1960 from New York City where he earned a master's degree from Columbia University. He has focused sharply on building technology throughout his career, learning much as a participant in Federal Emergency Management Agency (FEMA) programs on nuclear, wind and earthquake hazards. In the mid-1970s he moved to Florida, where he assisted in organizing the Florida Solar Energy Center, following which he returned to Utah as Executive Director of the state's seismic safety program. When the earthquake project was completed four years later, he and Crawley formed a partnership, Structural Facilities, Inc. He also returned at that time to the University of Utah's Graduate School of Architecture, where he was an adjunct associate professor until retiring in 1996. Ward is a long-time member of the Earthquake Engineering Research Institute (EERI) and served on its board of directors from 1985 to 1987.

Crawley is principal author of the textbook, *Steel Buildings: Analysis and Design*, recognized as a standard on the subject and now in its fourth edition. He came to the University of Utah in 1958 with a master's degree in architecture from Rensselaer Polytechnic Institute. He retired in 1995 as professor emeritus after 37 years. In 1993, Crawley received the Distinguished Professor Award from the Association of Collegiate Schools of Architecture, in recognition of his career as an outstanding teacher of structures. Crawley, too, expanded his knowledge of building technology through FEMA programs, and also as a structural consultant to architects and engineers in the Salt Lake City area. He is a member of the Structural Engineers Association of Utah, of which he has been a director. He was also appointed a member of the governing board of Utah's earthquake safety program and served for the lifetime of that program.

ACKNOWLEDGEMENT

Recognition is given to those who reviewed early drafts of this manuscript. Helpful comments came from James Ambrose, author of several well-known structural text books; Richard Chylinski, FAIA; Walter Lewis, Chair of NCARB's Continuing Professional Development Committee, Gabor Lorant, FAIA; Margarita Khuly, Florida Board of Architecture; and Billy R. Manning, who, at the time, was with the Southern Building Code Congress International (SBCCI). The authors are grateful for their suggestions.

FOREWORD

A Manual For Architects

This monograph provides information about wind and its implications for buildings, and limits mathematics only to determination of design wind pressures for illustrative examples. It must be noted that wind behavior and its effects are governed by physical principles most readily described mathematically rather than conceptually. Moreover, building resistance to wind forces must ultimately be treated numerically.

Accordingly, the authors have presented basic principles and concepts as a discourse, but have not hesitated to include mathematical equations when needed or when ideas are more fully described in their mathematical context. Still, the monograph is not an engineering text on wind.

While an architect may defer much of the quantitative work involving wind forces to his or her engineering consultant, it is not always feasible to do so for the myriad details produced by the architect in construction drawings. Many of those details involve the connection of one component to another–components that individually may experience wind effects.

It seems sufficient at this point simply to observe that wind loadings on building surfaces must somehow be resolved in the overall structural system, and that building components must be joined to other components with sufficient strength to prevent blowing away or tearing loose. An architect is responsible for the performance of the entire building under the standard of care mandated by licensing.

Every major storm and hurricane produces building failures–sometimes total collapses, sometimes only roof damage or broken windows–which demonstrate inadequate wind resistance. One possible cause of such failures is wind producing unexpectedly high pressures.

Another, however, is that connections at one or more locations were designed and/or detailed below accepted minimum standards. While an architect cannot alter the unexpectedly high wind pressures, connection details and their strength certainly are within the province of an architect's control.

The ultimate goal of this monograph is to provide information that will help produce wind-resistant buildings. Much of the information will serve to call the architect's attention to locations of vulnerable components. Some of the information requires an understanding of wind flow and how and where pressures develop on buildings. And, of course, some aspects require force or other kinds of calculations in order that the required strength be provided.

It also should be stated that design to resist wind forces, in practice, is rarely separated into distinct topics, as is done in the discussion that follows. Rather, concepts, considerations and numerical procedures for wind typically are amalgamated in building codes and standards.

These codes or standards usually include both prescriptive requirements and performance requirements, and often the two are mixed. One consequence of this amalgamation is that wind behavior, pressure development and building response often are not self-evident. Reliance solely upon meeting code requirements as the basis for treatment of wind provides little opportunity for exercising design judgment or, for that matter, and more importantly, for developing judgment.

This monograph aims at expanding the architect's understanding of wind forces as they apply to buildings so that professional services can be rendered by and for the design team with the requisite accountability.

The sequence of discussions in this monograph aims first at introducing fundamental concepts as well as associated vocabulary and definitions. Step by step, a technical basis is developed which allows examination of building behavior when subjected to wind.

Once these concepts are treated, structural aspects of building behavior will follow. Structural behavior is explained both in terms of structural concepts and the forces induced by wind which must be resisted. Next, construction practices, as established by building codes and standards, will be discussed. Finally and most importantly, construction practices applicable for new and retrofitted existing buildings will be considered, since these respond to the practical need of architects.

Wind—The Good and the Bad

The earth has an atmosphere that both permits and sustains life. The earth is in continual rotation and elliptical motion, with its poles changing distance from the sun, giving us seasons of the year. Consequently, the earth's atmosphere is moving and changing. Movement of air can enhance our well-being or endanger our lives. Mild movement of air is usually refreshing and healthy, and architects frequently capture this characteristic and integrate it in their design solutions. Sustained winds have even been utilized to generate electric power on a scale larger than the traditional farm—a modern technology illustrated in Photo 1.

On the other hand, more rapid and violent movement of air may be harmful–particularly to buildings. Such damage may be limited to broken glass or torn roofing membranes, or more extensive structural damage might occur, including total destruction in some instances. Whatever its extent, the effects are costly and disruptive for building users and owners.

Photo 1

Photo 2

Photo 3

All photos courtesy of the Federal Emergency Management Agency

A wall of the building shown in Photo 2 was seriously damaged as Hurricane Andrew swept across Florida in 1992. Although serious, this damage was far less than that shown in Photo 3 caused by the same hurricane winds.

Hurricane Andrew (Florida, 1992) and Hurricane Iniki (Hawaii, 1992) forcefully demonstrate a building's vulnerability to extreme winds. The insurance industry's Property Claim Services reported an estimated $23 billion insured loss in 1992, with 95 percent of that amount caused by wind and hail. With losses of this magnitude, largely due to two events, it is not surprising that building standards and practices involving wind resistance are currently receiving increased attention. In conjunction with this increased attention, the insurance industry, in cooperation with model code organizations, has developed a "code enforcement grading system" to rate each community on its code adoption and effectiveness–presumably for the purpose of applying an insurance rating system similar to the fire-protection grading system that has long been in use.

The building industry also has responded to these and other recent disastrous events. There is increasing concern regarding the suitability of construction practices as well as greater diligence both in proper design and at-site inspections. Refinements and changes in wind codes and standards are occurring today at an accelerated pace and are addressed throughout this monograph.

Effects on Architectural Practice

The practice of architecture has undergone fundamental change during the last half of the twentieth century. Technology and societal changes are the root cause.

First, there was a gradual shift from providing full architectural services within an office to the widespread use of consultants–specialists in structural design, mechanical systems, geotechnical, acoustics and construction management.

Advances in technology lead to increasingly complex design analysis, expanded code requirements, expanded liability (if not in concept, certainly in accountability) and greater expectations by the clients regarding building performance.

Social changes arising out of concern for the disabled, concern for the environment and excessive waste of raw materials and energy are now standard concerns in architectural practice.

Most recently, evolving use of the computer with its graphics capability has rendered the traditional draftsman almost obsolete.

The practice of architecture also has been affected by the cascading effects of regional disasters, notably hurricanes and earthquakes. Among the effects are more stringent design requirements, increased accountability regarding building designs, greater expectations regarding building performance and more extensive and rigorous code requirements.

All of these social and technological considerations, many of them beyond the individual architect's control, have incrementally expanded the nature of architectural services. Yet, while more specialized personnel are needed to meet the increasing demands placed on the architect, and while design documents are expanded in scope and detail, the profession has held to the notion that the architect assumes, by fact of licensing, full responsibility for building performance as required by the applicable building code. Each new occurrence of disaster or building distress tends to intensify user expectations or governmental requirements regarding building performance.

It is a fact that the architect is being held increasingly accountable for proper design to resist wind forces. Such accountability is manifest in the underlying contractual and regulatory basis for architectural practice–namely, the protection of life, safeguard against injury and reduced property damage.

Theoretically, it is possible to design nearly perfect wind-resistant buildings. A complete load path, adequate material strength and strong connections often are sufficient for ordinary low-rise buildings. But other not so ordinary buildings, including tall buildings and those subjected to special wind conditions, may require more sophisticated design solutions. For example, taller, wind-sensitive buildings have benefitted from recent technical advances involving mass dampers. Tornado-resistant buildings are even possible if one accepts a fortress-like environment. Theory, however, is quickly replaced by practicality, when cost enters into the discussion. Livability and functional suitability may also prove to be incompatible with a fortress-like environment.

Accordingly, the definition of an injury-free, undamaged building under wind forces must be placed in proper context. The wind-force requirements of the model building codes establish minimum design standards believed by experts to be sufficient in most, but not all, wind conditions.

For example, design for tornadoes is not considered in current codes, but hurricane-intensity winds are considered, although not at their full potential and not in every region. Buildings typically are designed for extreme winds in their local or regional environment based upon mathematical probabilities drawn from wind histories.

Basis for Practice

There are differences within the three widely used model building codes in the United States–National Building Code (BOCA), Standard Building Code (SBCCI) and Uniform Building Code (ICBO)–with respect to their wind provisions. The wind provisions in each are based upon fundamental principles; the differences are mainly matters of procedure resulting from the manner of assembly of equations and data, although there are also differences in degree of rigor required for the wind force analysis and in some minimum requirements.

There is a broadly accepted standard on wind forces for building design, now available from the American Society of Civil Engineers (ASCE) but formerly promulgated by the American National Standards Institute (ANSI), namely, *Minimum Design Loads for Buildings and Other Structures*, here referred to as ASCE 7. The ASCE 7 standard is considered to be the base reference today for the model code organizations. Each has adopted its provisions in whole, substantially, or with modifications (usually by rearranging and/or revising data tables and associated functional equations).

Since there are differences in procedures of application among the model codes, the portions of this monograph dealing with applications follow procedures of the ASCE 7 base reference. Significant differences between the ASCE procedures and those found in the three current model codes are identified, but the variations are not treated in the same depth. Of course, the applicable procedures are to be used when a particular model code is followed. In order to present a national point of view, the authors have chosen to present fundamental information in as general a way as feasible.

Preparation of the monograph has been influenced by another consideration. The most recent ASCE 7 standard for wind loading, ASCE 7-95, alters some long-standing procedures for establishing wind pressures on a building. The authors have elected to prepare this monograph based on procedures of ASCE 7-93, which is the standard most familiar to architects, and to identify and discuss issues treated differently in ASCE 7-95. Significant changes will be pointed out in the interests of alerting architects to evolving developments.

This leads to a concluding remark on the topic of building codes. The monograph does not attempt to discuss individual distinctions among the model codes regarding wind forces. A separate manual could be written treating the details of procedures for each model code. Since comprehensive treatment of individual codes is not an objective of this monograph, either ASCE 7-93 or ASCE 7-95 will be the usual reference for discussions when greater specificity is necessary.

WINDS AND BUILDINGS

Wind Considerations For Buildings

Throughout this monograph, concepts involving wind and buildings are illustrated by means of figures in which the wind being considered is shown to be blowing in a particular direction. We emphasize here that the wind direction shown in the figures is purely for purposes of illustrating concepts.

In fact, wind may blow in any direction. Although statistical analysis of wind histories might suggest a prevailing direction of peak velocities, it is not considered germane in the design of buildings. Rather, it is accepted procedure that winds acting upon buildings are considered to blow in any direction. Accordingly, for building applications the design wind velocity is applied in each direction on each of the building's orthogonal axes and a wind force analysis is completed for each condition. Building resistance is evaluated for the worst-case forces obtained from each of these analyses acting in combination with forces induced by portions of the gravity load (codes and standards typically specify the loading combinations that are to be considered).

It will be seen later that while wind typically is assumed to blow in a horizontal direction, pressures are developed having both horizontal and vertical components. Hence, building resistance must be developed for all of the wind force components. It is not enough to treat only the lateral forces.

Winds change constantly, both in velocity and in direction. Changes over short time intervals may produce a dynamic effect on buildings. For most projects, however, a dynamic analysis of wind effects is neither needed nor done—an exception being tall, slender buildings that are sensitive to rapid wind fluctuations. For nearly all other buildings, the accepted method of analysis is to convert wind forces to an equivalent static effect. Static procedures are the focus here, for they have wider application for architects. Dynamic evaluations involve considerations beyond the scope of this monograph.

Regional Variations of Wind

In general, winds of some degree of intensity affect all buildings, and resultant forces therefore require consideration for all buildings. Certain winds, notably those associated with hurricanes and tornadoes, are given names; whereas those extreme winds of broader interest are usually identified simply as wind storms. For this monograph, these winds are called general winds or, more often, just winds.

Velocity, or wind speed, is a characteristic common to all winds. In the context of building design for wind forces, velocity typically is the way winds are described and is a critical factor for wind effects. Velocity, a rate of change measured in miles per hour (mph), is sometimes associated with a particular type of wind; for example, wind velocity of a storm must reach 75 mph before it is classified as a hurricane.

Regional variations of winds often are distinguished in the United States by their names. Hurricanes, we know, produce great wind speeds over very wide areas. Wind velocities in excess of 140 mph have occurred, yet these regional phenomena are commonly associated with the southeastern and Atlantic Coast states in the continental United States, and with Hawaii. Thus, when we speak of extreme winds in those areas, we mean winds of hurricane velocity, which is the basis for defining wind velocities that are used for building designs in those areas.

Tornadoes, too, occur much more frequently and more violently in some regions than in others, although hardly any state is free of them. Central and southern states in the United States experience tornadoes most frequently, but these are not necessarily more destructive than the less frequent tornadoes elsewhere. This is because the principal determinants of destructive effects —namely, extent of ground path and dynamic wind effects, along with extreme velocities — do not appear to be associated with any particular region. Tornadoes, with wind velocities of 200 mph or more, can be more extreme than hurricanes, but typically they affect much smaller areas. Tornadoes also are distinguished from other wind storms by their abrupt loading effects (a dynamic loading consideration) and by large differences in pressures as the tornado passes, which sometimes results in explosion of closed buildings. Tornadoes occur more frequently than hurricanes, a fact that would need to be considered if tornado-wind velocities were included when establishing design velocity for a building. However, buildings normally are not designed to resist tornadoes, and procedures in codes and standards do not consider them or their dynamic effects.

The architect's ultimate concern is to provide an acceptable level of building resistance to wind forces. The decision as to what constitutes an acceptable level of resistance is a socio-political decision, nowadays largely decided by states and municipal governments through the ordinances and building codes they adopt. The basis for a given jurisdiction's decision is its proximate wind velocity, as established from wind histories. Thus, regional differences in peak wind speeds are identified and provide the basis for giving meaning to the term "acceptable."

Because of the widespread impact of hurricanes and economically reasonable construction practices to resist their effects, hurricane wind velocities have been included in codes and standards. Conversely, tornado wind velocities are not reflected in codes and standards because of their localized impacts and extreme velocities for which resistant construction, mandated broadly, is neither economically nor practically feasible.

Wind velocities as the basis for building design typically have been mapped using velocity contours to denote variations. Several different wind velocity maps have been developed, and velocity contours sometimes are slightly different; for example, the wind map in the 1982 Standard Building Code (SBC) differs from the map in the 1982 Uniform Building Code (UBC). Differences largely depend upon the publication date of the original (source) map, inasmuch as the source for most wind map data is the U.S. Weather Service.

However, differences in the maps may also be due to what is being mapped. Until recently, most maps were based upon fastest-mile wind (defined in the section on "Wind Records") in open terrain at a height of 33 feet. Another, based upon three-second gust velocity (also defined in "Wind Records") and for open terrain and a height of 33 feet, has been introduced in ASCE 7-95 and may be adopted in model codes as a replacement for today's fastest-mile wind maps. Mapping of tornado-velocity winds also can be found in the literature; ASCE 7-95, for example, includes such a map in the Commentary.

Maps shown in ASCE 7-93 and ASCE 7-95 illustrate a fastest-mile wind speed map and three-second gust wind speed map, respectively. As indicated by the titles for these maps, wind velocity is called basic wind speed. This term assumes a specific connotation when building wind pressures are being determined.

The maps also indicate a specific height above the ground and "roughness" conditions of the terrain for which the mapped velocities apply. This information, too, is used later when pressures are determined. For a particular building application, the velocity map in the operative building code, or as prescribed by the building authority of a jurisdiction, must be used as a minimum.

Architects should note that velocity contours appearing in the fastest-mile wind speed map and the three-second gust wind speed map are different. This has significance beyond the fact that larger velocities appear in the three-second gust wind speed map. Later discussions will explain how applications procedures rely largely upon data tables that have been compiled from one or the other of the two map types.

Because the mapped velocities are based upon different criteria, data tables are not interchangeable. Everyone must exercise caution using maps and tables, especially as the three-second gust wind speed map and associated procedures become widely adopted.

THE NATURE OF WIND

Theoretical Foundations

Wind flow and its interaction with buildings can be described with a few principles from classical physics, and procedures for evaluating this interaction are founded upon these principles. The principles pertain to flow of air, interference with flow and development of pressures.

A better understanding of the theoretical foundations of air flow enhances one's appreciation of the practical approach taken in evaluating wind interaction with buildings. Practical procedures rely heavily upon working equations and data from tables which often obscure the fundamentals.

In this monograph, our interest is in advancing intuitive knowledge rather than with number crunching, so emphasis is given to concepts.

The Physics of Pressure

Pressure is defined simply as force per unit area–with units being pounds per square inch (psi) or, as is more convenient for discussions of wind, pounds per square foot (psf). Hence, pressure and force sometimes will appear to be used interchangeably since in a mathematical sense both are vectors. To keep possible confusion to a minimum, the term pressure (psf) will be used in discussions involving development of the wind loading on building surfaces. Force (lbs) will be used when the pressures have been aggregated, such as a resultant of pressure over a defined area, or possibly a distributed force where the pressure resultant has been aggregated on a particular building component.

From physics, we know that pressure always is applied perpendicular to (normal to) a surface. This is so regardless of the cause of the pressure–whether from air flow, flow of other gases, fluid flow or due to a pressure differential between two surfaces.

Pressure is a consequence of (1) molecules of a fluid, gas, or air impacting a surface; (2) currents created by flow of air, gas, or fluid around or across a surface; or (3) a pressure differential on opposite sides of a surface. Each of these can be a consequence of wind flow.

Mathematical Description of Pressure

The mathematical description of wind pressure is simple to state and easily remembered–much easier to understand and apply than is the mathematical derivation which draws upon a combination of several principles of physics. The familiar expression is:

$$q = 0.00256\ V^2$$

The q in this expression represents a very particular pressure called the stagnation pressure. This pressure is discussed in detail after a brief discussion of the physics of air flow. The V represents the velocity (speed) of the wind in miles per hour. The coefficient 0.00256 includes the necessary conversions so that q has units of pounds per square foot. The calculated wind pressure on a building, commonly called p, is not the same as q, and the values likely are not the same, yet the stagnation pressure is needed to obtain p. Pressures calculated using this equation for q should not be confused with the building design pressures. The meaning and significance of q become more understandable by examining the basic principles of physics from which it is derived. These principles are outlined next.

Wind is our everyday description of moving air. Air is comprised of molecules of a variety of gases. Wind, then, consists of a moving mass of loosely packed molecules called air. The moving air mass has, as described in common physics, kinetic energy; that is, it has energy capable of doing work. The mathematical formulation of this kinetic energy is $k = \frac{1}{2}mv^2$ where m is mass and v is the velocity of the mass. Units for k are ft-lbs; units for m are w/g, or weight (lbs) divided by gravity (g, or ft/sec.^{2}); units for v as used here are ft/sec.

q = stagnation pressure
v = velocity (speed)
p = calculated wind pressure
k = kinetic energy
m = mass

This definition of wind energy is used in combination with a mathematical formulation of fluid flow, a concept of physics defined by Bernoulli's Equation. Bernoulli formulated a fundamental equation of hydrodynamics which provides an important relationship between pressure, velocity of flow, and points along the flow path.

Fluid Flow and Pressure

Fluid flow can be thought of as a flow stream sometimes described as a number of parallel filaments. When stream flow is important to a discussion, wind sometimes is represented graphically as a group of parallel lines, with arrows to indicate the direction of flow, as appears in Figure 1a. This method of representing wind flow is used in initial discussions in this monograph as a way to describe how and where pressures develop on building surfaces.

Direction of flow is an essential aspect of pressure–how pressure develops and the direction of the pressure vector. Accordingly, a flow direction is indicated in ensuing diagrams involving wind pressure. In many situations graphical representation of stream flow is unnecessary, and only the direction of flow is needed. Figure 1b illustrates the graphical representation used later in this monograph when only the direction of the wind is needed.

Bernoulli's Equation Relating Pressure and Flow

Bernoulli's Equation states that there is a very simple relationship between fluid (also air) pressure and velocity measured at one point and the pressure and velocity measured at another point along the path of stream flow–specifically that pressure plus kinetic energy of the fluid at the first point equals pressure plus kinetic energy of the fluid at the second point. This relationship is described with the equation

$$p_1 + \frac{1}{2}mv_1^2 = p_2 + \frac{1}{2}mv_2^2$$

and can be interpreted to mean that pressure plus kinetic energy is a constant value for a particular fluid and flow. The equation derives from definitions of work and energy.

Note that kinetic energy is determined by the mass density and velocity of the fluid. Hence, for example, if velocity at the second point increases, then there must be a corresponding decrease in pressure at the second point of the same fluid so that the constant value remains unchanged. The significance of this equation is that if one of the variables of pressure or velocity is changed, then there is a corresponding change in the other variable so that the result remains the same; that is, the value of combined terms on each side of the equal sign remains constant.

The Bernoulli effect is illustrated in Figure 2 for

Figure 1a AIR STREAM FLOW

Figure 1b WIND DIRECTION

fluid flow constrained in a pipe of different diameters along its length. A volume of fluid at a pressure p_1 passing between two points A and B on the left side over a particular period of time necessarily must push the same volume of fluid through the pipe of smaller diameter on the right, represented by points A' and B' in the figure. Fluid velocity on the right side necessarily is increased. Mathematical derivation of the Bernoulli Equation shows that this occurs with a reduced pressure p_2 corresponding to the increase in velocity.

Air Flow

The same concepts and principles apply when considering air flow past an obstruction. Consider the flow of air having a very wide stream past an object of smaller dimension, illustrated in Figure 3. In this case, the stream lines flow around the obstruction. Those stream lines nearer the obstruction necessarily have greater velocity so that the same volume of air flow is maintained. The stream flow gradually evens out at some distance away from the obstruction, and the stream lines eventually return to their unobstructed state. This diagram is intended only to show the diversion of the overall stream flow and neglects other significant effects on windward, leeward and side surfaces which are discussed in detail in the section on pressures.

Even with this incomplete diagram of Figure 3, the Bernoulli effect is implied. At the windward surface, stream flow velocity is slowed as the air stream moves around the obstruction. In accordance with Bernoulli's Equation, the decreased velocity must be accompanied by a pressure increase over ambient pressure of the unaltered air stream in order to maintain the constant relationship. Such an increase over the ambient pressure is called positive pressure and acts perpendicular to and toward the windward surface.

Because of the increased flow velocity near the sides of the obstruction, there also occurs a pressure change on the side surfaces in accordance with Bernoulli's Principle. As the stream flow velocity increases, there is a corresponding drop in pressure on the side surfaces of the obstruction. Such a reduction in pressure on the side walls is called negative pressure and acts in a direction still perpendicular to, but away from, each of the side surfaces.

Stagnation Pressure

That surface of the obstruction perpendicular to and

Figure 2 BERNOULLI EFFECT ILLUSTRATION

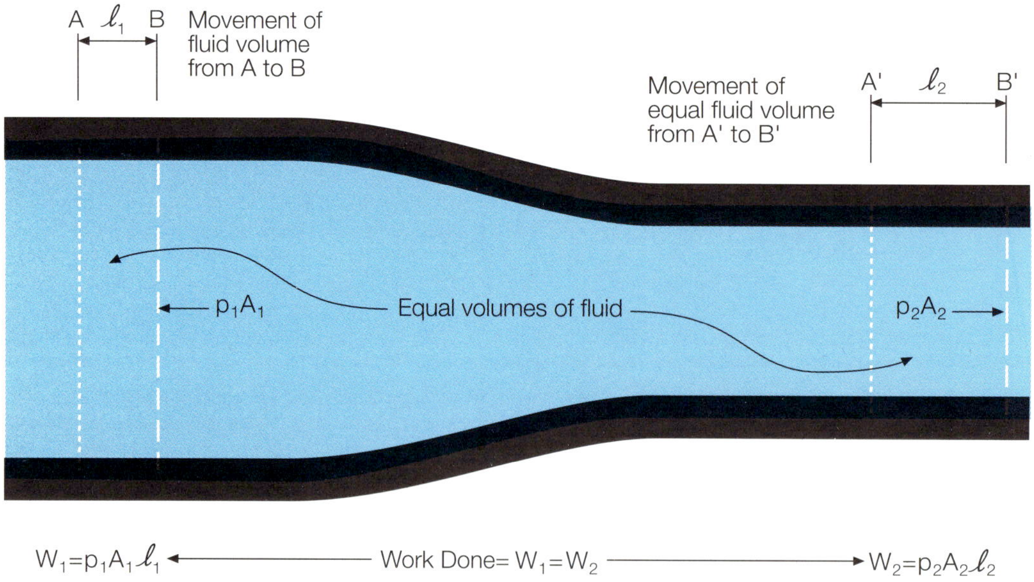

Equal work done at two locations of fluid movement

facing the direction of flow provides the conditions for defining the stagnation pressure. If that facing surface (called the windward face in later discussions) extends indefinitely, both laterally and vertically, so that the wind stream is completely blocked–that is, has zero velocity as shown in Figure 4–the pressure acting on that facing surface is the kinetic energy, or simply ½ times the total oncoming flow times the velocity squared. This value is the stagnation pressure, *q*, and represents the largest pressure that can occur on the obstruction. This pressure will act in the direction of flow.

The specific coefficient of 0.00256 given at the outset of this discussion is obtained from values for the density of air and correlation of units for mass and velocity. Clearly, air density is not a single value. Moisture content and height above sea level produce varying mass densities for air. These variations are not taken into consideration for the coefficient of 0.00256, and a fixed air density typically is used in wind pressure calculations. The standard density taken for air is 0.07651 pounds per cubic foot at 15° C and at 760 mm mercury pressure (approximately sea level). Wind velocity usually is in miles per hour. When the standard density is introduced into the equation and conversion of units is undertaken, the stagnation pressure value obtained is

$$q = \tfrac{1}{2}\,(0.07651 / 32.2)\,(5280 / 3600)\,v^2 = 0.00256\,v^2$$

Units of measurement are

$$q = (\text{lb/ft}^3 / \text{ft/sec}^2)\,(\text{ft/mile} / \text{sec/hour})\,(\text{mile/hour}^2) = \text{lb/ft}^2$$

Figure 3 BERNOULLI EFFECT IN AIR FLOW

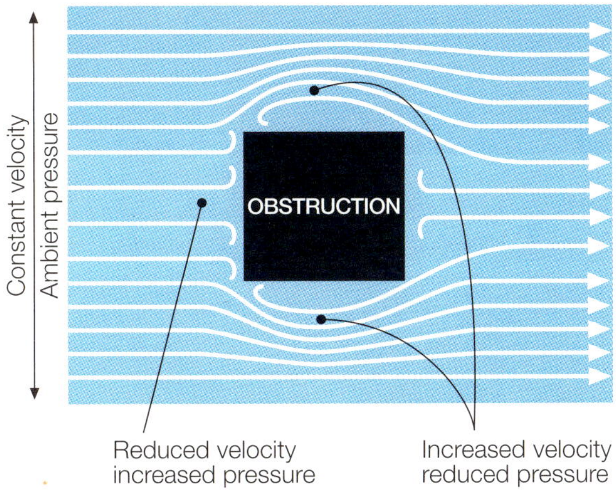

Information is available for other air densities, and the coefficient of 0.00256 would be different if some other density is used. Such use might be warranted, though infrequently, for conditions of extremely moist air and when more precise wind pressures are desired, as might be the case for a wind-sensitive building in a hurricane region.

Stagnation pressure has been defined as that pressure at locations of zero velocity occurring when an obstruction extends infinitely across the flow path. Buildings have specific dimensions; hence *q* is not actual pressure but is used to obtain actual pressure, or more particularly to obtain a pressure used for building design purposes.

Factors Affecting Velocity of Flow

Additional characteristics of wind flow have direct application to pressures developed on buildings.

The previous discussion of fluid flow revealed that fluid pressure and velocity are interrelated. Pressure influences velocity in constrained conditions (such as within a pipe), and velocity influences pressure when confinement constraints are removed or altered. Wind velocity and pressure follow these principles in the context of ground surface conditions and

Figure 4 STAGNATION PRESSURE ILLUSTRATION

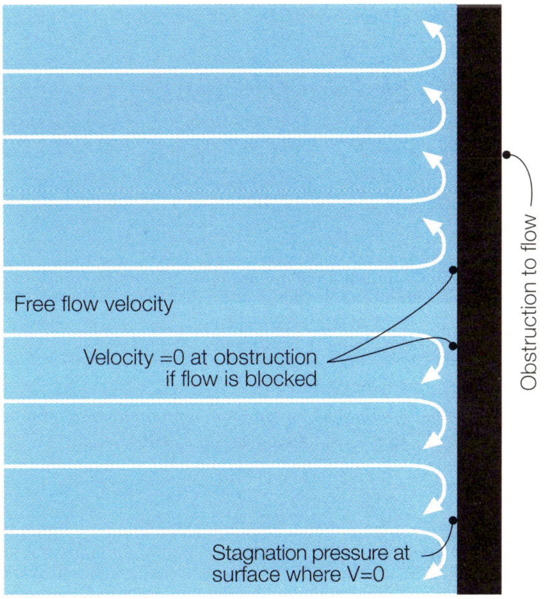

natural topographic features, as well as man-made obstructions. The general concepts governing interference of free flow are examined next. Then, the concepts are linked to wind velocity.

Boundary Layer

Not only is air flow (the air stream) deflected as it flows around an obstruction, the flow also is retarded as it moves past a surface which is parallel with the path of motion. This retardation is caused by friction at the surface which interferes with smooth flow of the air.

An important first observation in this regard is that the air velocity at the surface of the parallel boundary must be zero. The flow velocity then increases away from the boundary. The retardation effect, which is maximum at the surface, is transferred perpendicular to the surface into the neighboring air stream by fluid viscosity. This is a characteristic of any fluid and relates to molecular adhesion within the fluid.

This characteristic of all fluids and gases once again can be described most easily using the previous example of fluid flowing through a pipe. This time, a pipe of constant diameter is used to demonstrate the effect. However, for this example, no attempt is made to provide a mathematical model of the behavior. A mathematical description will be applied later to wind in which the ground surface is analogous to the bottom side of the pipe but where there is no top surface in the atmosphere.

Fluid flowing through a pipe with smooth walls is illustrated in Figure 5a. If the velocity of flow is somehow measured at different points over a cross section, it is found that the velocity varies from zero at the surface (boundary), increases to a maximum value at the center point of the cross section, then decreases once again to zero at the opposite side. Figure 5b shows a plot of the velocity over the cross section, commonly called a velocity profile. Increased roughness of the side walls of the pipe will produce lower velocities than those shown, but the general form of the velocity profile will be similar. For a pipe of larger diameter, the maximum velocity will increase slightly, and if the pipe diameter is made large enough an upper limit on the maximum velocity will be reached. This upper limit occurs when the effect of the surface friction no longer influences the fluid viscosity. The region between the surface point of zero velocity and the upper velocity limit is called the boundary layer, and the effect is called the boundary layer effect.

The importance of this characteristic of fluid flow to wind engineering applications is that the surface of the earth acts as a restraint on the wind. There exists above the earth's surface a boundary layer where the retardation on wind flow is evident. This is the explanation for

Figure 5a STREAM FLOW

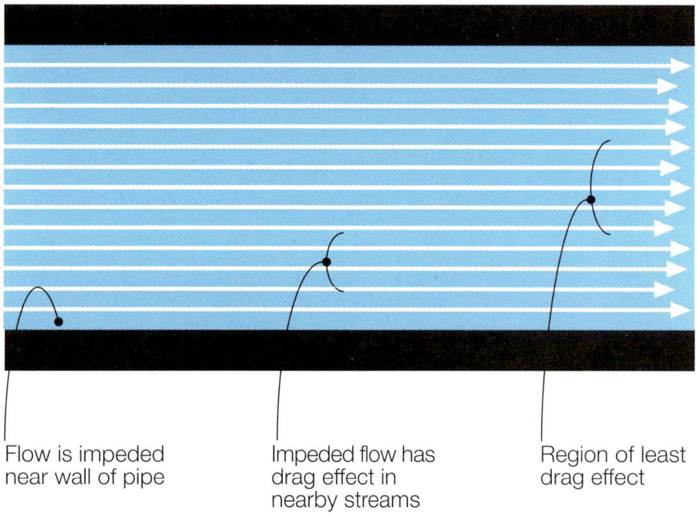

Figure 5b VELOCITY PROFILE

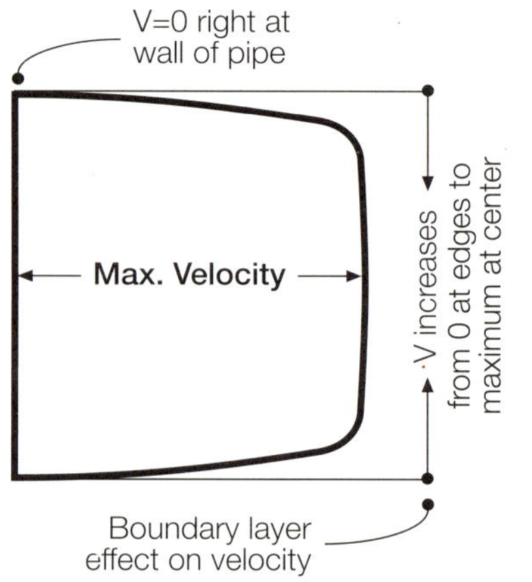

the commonly heard statement, "Wind speed varies with height."

Because there are many degrees of roughness at the earth's surface, the boundary layer has variable height above the surface. This variable also is considered when wind pressures are determined.

Gradient Height

The boundary layer influence for wind is most meaningful when examined on a large scale. Indeed, the boundary layer for wind effects must be examined at such scale.

The earth's surface has a wide range of surface roughnesses–the nearly flat surfaces of the oceans at one extreme to jutting mountains at the other extreme. In between are varying degrees of roughness, such as sparsely vegetated land and forests. Towns and cities add to the variety of surface roughnesses. Urban cores of large cities, suburbs and rural towns provide the range. Like the example above of fluid flow in pipes of different side wall roughnesses, the height of the boundary layer effect for wind varies in accordance with the degree of ground roughness; it is the same effect as looking at the lower half of the velocity profile shown in the pipe example. This means that the height above the ground surface at which the upper limit for wind velocity is reached is dependent upon the surface roughness.

The gradient height is that elevation above which the friction influence of the earth's surface is no longer significant in retarding air flow. Above this gradient height, wind velocity is assumed to be a constant maximum value for a given wind, called the gradient velocity. Below the gradient height, different velocities will occur at different heights for the same wind. The gradient height varies as surface roughness varies. In general, it increases as surface roughness increases.

Characterization of the Boundary Layer Influence

The gradient height varies due to several additional complex factors, including pressure gradient of lower and upper atmosphere, coefficient of eddy viscosity of the air, mass density of the air, angular velocity of the earth's rotation, geographic latitude and curvature of the wind path. Several mathematical models have been advanced to characterize this variation of wind velocity as a function of height above the earth's surface. Only one of these models is treated in this text, the model upon which current procedures are based for defining wind velocity for engineering characterizations, called the power law. The other models are of use primarily to meteorologists.

Power Law

Current building codes and standards for wind are based upon the power law, which is an exponential mathematical model. This model is almost universally accepted for architecture and engineering applications. It is used to describe wind-velocity variation with height. As will be seen later when numbers are attached to the model, nearly all buildings, including the tallest, fall within the boundary layer.

The general form of the power law as applied to wind is

$$\frac{V_Z}{V_X} = \left(\frac{Z}{X}\right)^{1/\alpha} \qquad \textbf{Equation 1}$$

where V_z is the wind velocity at height Z above the ground, α is a constant relating to ground surface conditions (to be discussed in greater detail), and V_x is the wind velocity at height X above the ground surface.

This equation provides a means for relating wind velocities, influenced by the same boundary condition, at two different heights. If this equation is rearranged so that V_x appears on the left side and a velocity at height Z is known, a velocity at any other height X within the boundary layer region can be determined. The parameter α is chosen so that the equation corresponds closely to an empirically established wind velocity profile for a particular ground surface roughness.

$$V_X = V_Z \left(\frac{X}{Z}\right)^{1/\alpha} \qquad \textbf{Equation 2}$$

For many applications to buildings, more specific forms of these equations often appear in relevant literature. Two particularly useful forms are provided here, because they are often used. One useful form of Equation 2 is

$$V_X = V_{33} \left(\frac{X}{33}\right)^{1/\alpha} \qquad \textbf{Equation 3}$$

and permits determination of a velocity V_x at any arbitrarily chosen height X based upon a known velocity V_{33} at a height of 33 feet. The 33-foot height, of course, is the basis for wind map velocities. The significance of the equation in this form will become evident when the exponent α is further explained.

$$V_z = \left[V_{33} \left(\frac{900}{33} \right)^{1/7} \right] \left(\frac{Z}{Z_g} \right)^{1/\alpha} \quad \textbf{Equation 4}$$

In this equation, 900 feet is the gradient height associated with a particular α (the terrain roughness discussed later); V_{33} is the velocity at a height of 33 feet; Z_g is a gradient height correlated with a particular α which value is 7; Z is that height correlated with another α for which V_z is calculated.

Equation 4 is obtained by mathematical manipulations (not provided here) so that velocity can be determined for one α condition based upon knowing a velocity at a specified height for another α condition. Again, the reason for wanting to do this will become evident as the discussion proceeds.

With Equations 3 and 4, nearly all wind-velocity conversions needed for building applications can be obtained. In codes, however, needed values have been calculated and placed in tables or integrated with other table values, so calculations using these equations are not often done.

Table 1 VALUES OF 1/α FOR DIFFERENT TERRAIN ROUGHNESSES*

DESCRIPTION OF TERRAIN	EXPONENT
Coastal Waters	1/10.5
Flat Shore	1/8.3
Open level farm land with isolated trees	1/7.8
Open grassland without tress	1/7.7
Flat, treeless grassland	1/7.4
Open slightly rolling farm land	1/7
Open grassland with few trees	1/7
Gently rolling grazing land with few trees	1/6.25
Open fields with low stone walls, hedges	1/5.9
Rough coast	1/5
Gently rolling country with bushes, trees	1/4.55
Level country with scrub oak and pine	1/3.3
Wooded farmland	1/2.8
Center of a large city	1/2

* Suggested by A.G. Davenport in an article published in ASCE Circular 3123.

Terrain Roughness

Table 1 shows values for α that appears in the power law equations for a variety of types of terrain conditions, i.e., surface roughnesses. The exponents are arranged in sequence beginning with relatively smooth terrain and progressing to increasingly rougher terrain–in other words, varying types of surface roughness which alter the boundary layer effect. The values shown for α are correlated with fastest-mile wind velocity records. They subsequently are modified to correlate with three-second gust velocities.

The work by Davenport which defines α for different terrain conditions provides the necessary bridge from theory to application. Actual wind-velocity data compiled in known conditions completes that information needed to apply the power law. This data comes from wind records of the National Weather Service in the United States. Typically, these wind records are compiled for a standard height of 33 feet. A value of 33 feet is used in standard maps and procedures of model codes and standards. Also, data in U.S. Weather Service records are produced either from measurements taken at a height of 33 feet or from values taken at another height and transformed to a corresponding 33-foot height using the power law with the value of α equal to 7. Note that the exponent value of 1/7 coincides with Davenport's terrain condition of "open, slightly rolling farm land" and "open grassland without trees" (see Table 1). This is terrain similar to that around airports, where weather stations typically are located.

Wind Velocity Profiles

It is useful at this point to examine wind velocity as described by the power law equation. The first issue to address is, what happens to wind velocity within the boundary layer as height above the surface increases? To demonstrate the overall effect, calculated values are plotted at the same scale in Figures 6a and 6b to show velocity at all height increments for two different terrain roughnesses. Each plot provides a wind-velocity profile.

Both velocity profiles shown in Figure 6 are based upon the same wind velocity of 100 mph at a height of 33 feet. This velocity is selected only to illustrate the boundary-layer effect for two different surface roughnesses. The value of α is taken as 3 (hence $1/\alpha = 0.33$) for Figure 6a and as 10 (hence $1/\alpha = 0.10$) for Figure 6b. These two values approximately bracket the range of α values provided by Davenport that were shown in Table 1.

Although calculated values are different, both velocity profiles have similar shapes. Both plots show a zero velocity at the surface and an increasingly steeper slope which approaches a straight vertical line at higher elevations. The value of the exponent merely alters the slope, not the general shape of the plot.

However, the height at which the plot approaches a vertical line also increases as α decreases. Figure 6a, which has the smaller α value (and also is representative of rough terrain such as the center of a large urban area), approaches maximum velocity at a greater height than occurs in Figure 6b, in which the value for α represents very smooth terrain.

That point where the plotted curve is essentially vertical is the gradient height earlier described. The wind velocity at gradient height is called the gradient velocity. The gradient velocity is reached at a lower elevation for smooth terrain than for rough terrain.

A comparison of the two diagrams of Figure 6

Figure 6a ROUGH TERRAIN

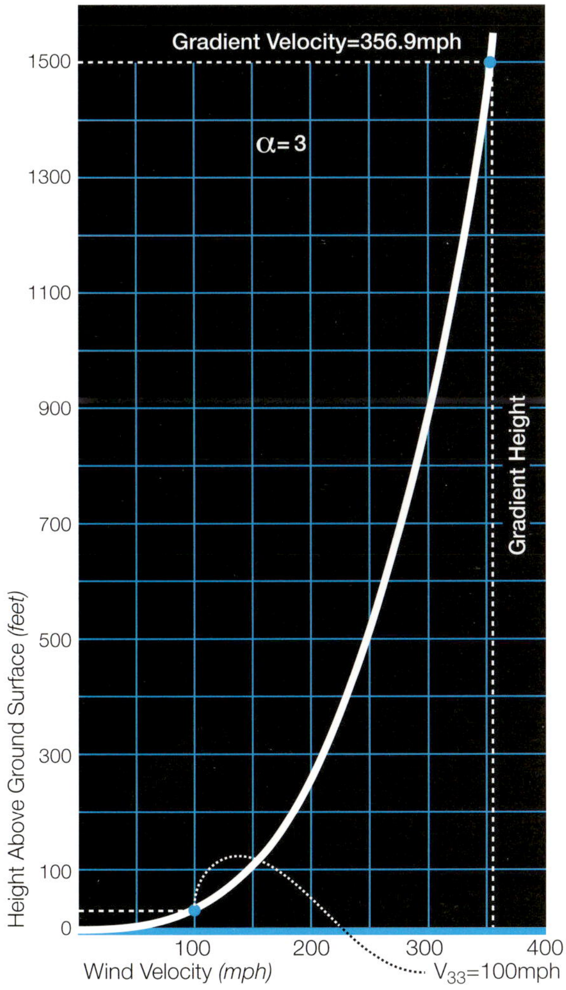

Figure 6b SMOOTH TERRAIN

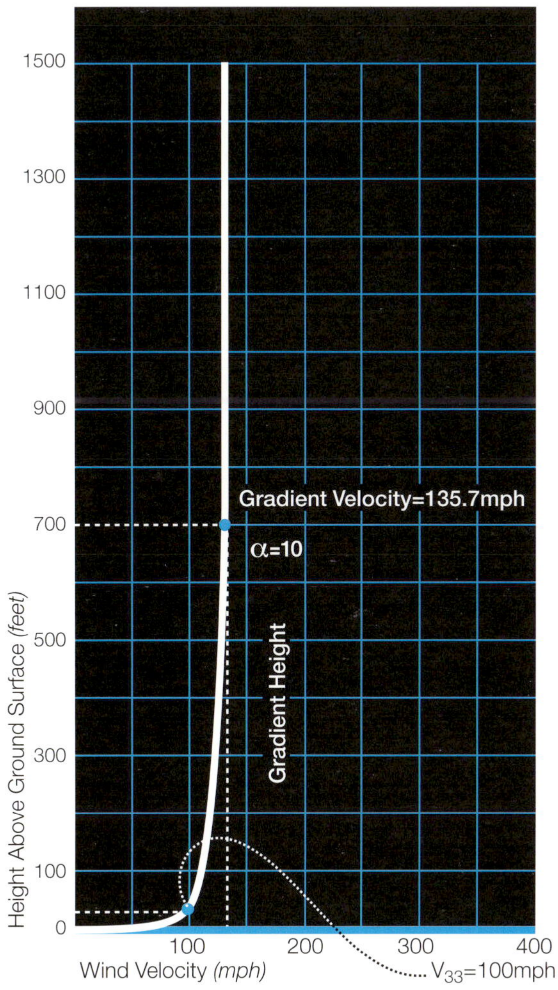

highlights yet another significant concept. Recall that both diagrams are based upon the same wind velocity of 100 mph at a height of 33 feet. Hence, both commonly would be called 100-mph winds. But, because terrain roughnesses are different, wind velocities at other heights are not the same. Indeed, there is a dramatic difference in the gradient velocities and gradient heights for the two diagrams. Even at lower heights, velocities differ (and differ enough) so that differences of pressure are further magnified when calculated using the square of the velocity.

The meaning of a particular wind velocity is further illustrated by the two comparable diagrams of Figure 7. The diagrams of this figure are based upon the same values of α as used in Figure 6, but the velocities are plotted based upon the same velocity of 100 mph at gradient height. It is observable that comparable velocities at a standard height of 33 feet are significantly different.

What, then, constitutes a 100-mph wind (the constant value in both diagrams of both figures)? It is accepted convention to define wind velocity at a standard height of 33 feet and, for application to buildings, for open terrain conditions–that is, for α = 7 when using fastest-mile wind data and α = 9.5 when using three-second gust data.

Figure 7a ROUGH TERRAIN

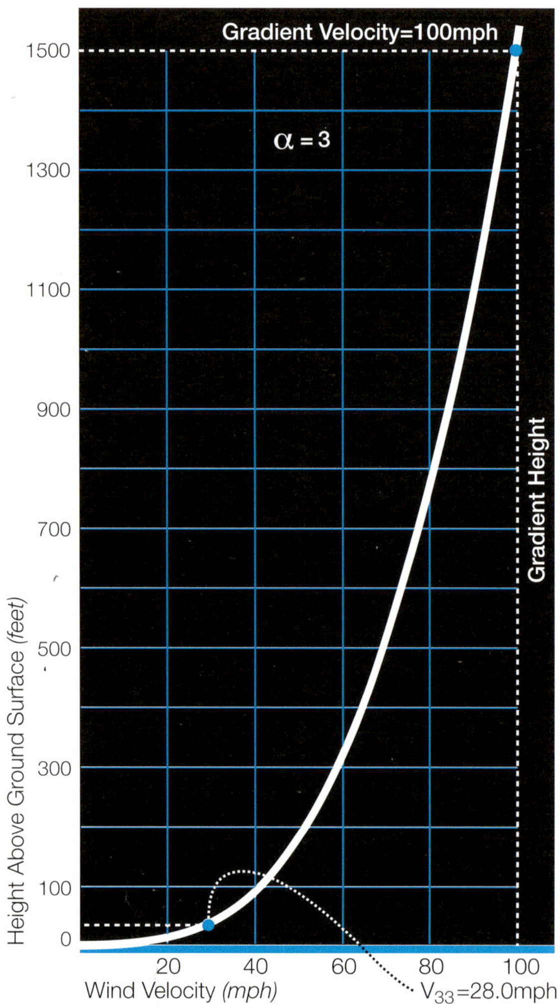

Figure 7b SMOOTH TERRAIN

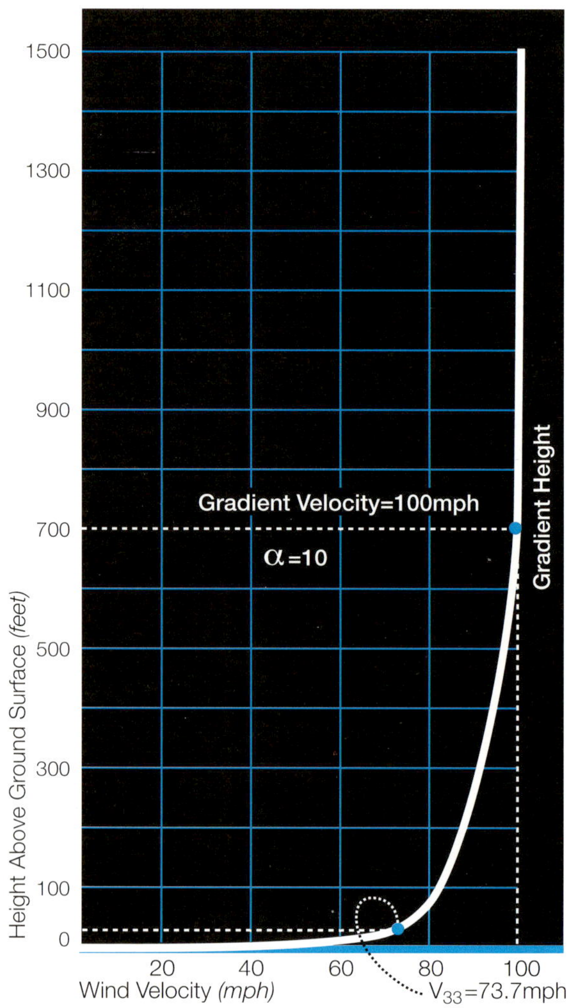

Velocity at any other height and for any other condition of surface roughness, then, is defined by the modified power law of Equation 4. This convention is essential in order to standardize any further discussion of wind and pressure.

Wind Velocity Profiles for Application

For architecture and engineering applications, a set of terrain conditions is used comprising fewer cases than provided by Davenport. A set of four terrain roughnesses has been accepted almost universally. The reduced set encompasses the range provided by Davenport but without as many values for a. The reduced set of four–their associated values for α, names by which they have become commonly known, and their associated gradient heights–are shown in Table 2 (fastest-mile wind) and Table 3 (three-second gust wind). The tables also list an Exposure Category for each of the four cases. Exposure Category is the usual manner by which terrain conditions are referenced for building applications.

Representative wind-velocity profiles for the four cases using fastest-mile wind data are illustrated in

Figure 8 FASTEST MILE WIND

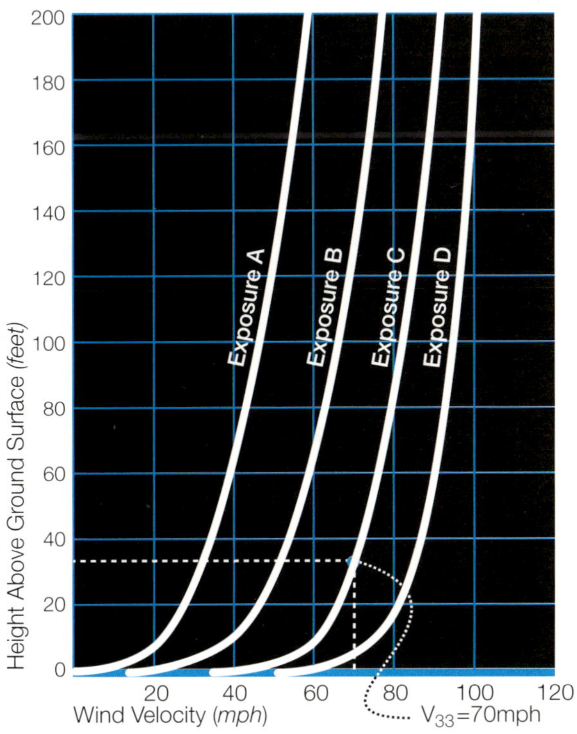

Table 2 STANDARD TERRAIN CONDITIONS/FASTEST-MILE WIND

TERRAIN CONDITIONS	EXPOSURE CATEGORY	α	GRADIENT HEIGHT (feet)
Urban Center	A	3.0	1500
Suburban	B	4.5	1200
Open	C	7.0	900
Coastal	D	10.0	700

Table 3 STANDARD TERRAIN CONDITIONS/3-SECOND GUST WIND

TERRAIN CONDITIONS	EXPOSURE CATEGORY	α	GRADIENT HEIGHT (feet)
Urban Center	A	5.0	1500
Suburban	B	7.0	1200
Open	C	9.5	900
Coastal	D	11.5	700

Table 4 WIND VELOCITIES (MPH)-DIFFERENT EXPOSURE CATEGORIES, DIFFERENT HEIGHTS

HEIGHT ABOVE GROUND (feet)	EXPOSURE CATEGORY			
	C Standard Exposure	A	B	D
20	93	38	64	112
33	**100**	**45**	**72**	**118**
50	106	52	79	123
100	117	65	92	132
200	129	82	108	141

Figure 8. All velocity profiles in the diagram, calculated from Equations 3 and 4, are based upon a wind velocity of 70 mph at a height of 33 feet for Exposure Category C (the velocity chosen is arbitrary and only for illustration). Velocities are plotted only to a height of 200 feet so that boundary-layer characterizations for the four cases are comparable at lower heights. Other choices for basic wind speed produce similar profiles but different velocity values.

The power law has several important uses. Used in its most general form, if a wind speed is known at a particular height (V_z and Z) and is influenced by known surface roughness (the effect of α), then the velocity for the same wind can be determined at any other height (V_x and X) for the same terrain. As shown in Figure 8, this is the use for Equation 3. Using Equation 4, similar determinations can be made for another surface roughness condition.

With these equations, one can take a velocity from a wind map (Exposure Category C, for a height of 33 feet) and convert that velocity to an equivalent velocity as would occur, say, in an urban location and for a 10-story building (Exposure Category A, for a height of 100 feet).

Because the exponential model becomes imprecise as gradient height is approached, the gradient height calculates less accurately using the power law. Accordingly, the gradient height typically is set in accordance with values provided in Table 2 or Table 3. This gradient height information is needed so that Equation 4 can be used to obtain equivalent velocities for other terrain roughness conditions. The gradient height of each Exposure Category must be known.

Table 4 provides comparative calculated values for a wind velocity as might be obtained from a fastest-mile wind map and converts that velocity, first to other heights in the same Exposure Category C, and second to equivalent velocities for the other standard terrain conditions. All values shown in Table 4 are obtained from the same basic wind speed at a height of 33 feet and are highlighted in the table. Architects may wish to verify values in Table 4 using Equations 3 and 4.

Velocities shown in the table column for Exposure Category C are calculated using Equation 3. Velocities in the table columns for Exposure Categories A, B and D are calculated using Equation 4. For example, the wind velocity at the top of the hypothetical 10-story building of 100-ft. height in an urban setting would be 65 mph (highlighted in the table). This is the applicable velocity based upon a mapped velocity of 100 mph, which is the velocity a designer would have established at the outset of a wind analysis.

These tabled values are for illustration only, but they are representative of numerical results for other basic wind speeds.

The four conditions of terrain roughness represent the underlying basis for all procedures to determine wind pressures in commonly used wind codes and standards. However, from a practical standpoint, computations of the sort just discussed are rarely, if at all, done by an architect. Instead, standard references typically provide the conversions in tabled form, although normally not in the form provided here. Data sometimes are presented in different ways, depending upon the reference. Tables sometimes provide wind profile values for selected basic wind velocities and for some, if not all, of the four conditions of terrain roughness. Table formats vary in different references. In some references intended for design applications, tables provide basic wind pressures for ranges of basic wind velocity, heights and conditions of terrain roughness, all in one tabled value for each basic wind speed. More technical references simply provide wind velocity maps from which velocities and pressures are to be calculated in accordance with applicable conditions.

If the various ways used in standard references to present essentially the same information get confusing, architects should simply return to the fundamental concepts as described in Table 4. The end goal, whatever simplifications appear in references, is to establish a wind-pressure profile in the applicable exposure category, which is the beginning information needed to establish wind loading on a building. Because most building heights fall within the lower elevations of the pressure profiles, a full pressure profile, which includes higher elevations, often is not needed. This, too, is a reason for preferring tables of data. It should be noted that, in practice, the profile need be extended no higher than the building height.

Review of Wind Pressure Concepts

The following summary may help to highlight important concepts regarding wind pressure in architectural applications:

- Wind velocity varies exponentially with the height above a surface.

- The degree of roughness of surface terrain has a retardation effect on wind velocity within the boundary layer (defined by the exponent).
- Rougher terrain has a deepening effect upon the height of the boundary layer above the surface.
- The heights of nearly all buildings lie within the boundary layer, and so those wind velocities impacting a building vary from zero at ground level to a maximum velocity which typically is less than the gradient velocity.

Wind Records

The discussion of wind pressure and velocity has provided both a theoretical and practical focus. Use of the mathematical relationships is dependent upon having available baseline wind-velocity data. Because wind varies in intensity over time at any given location, a standard basis is needed for input into the power law concepts. Such information is assembled from historical records of wind speeds. As noted previously, the U.S. Weather Service is the source of such information in the United States.

Wind speeds have been recorded over many years, mostly at airport locations, throughout the United States. The data from these records have been compiled in several forms. Two of these forms are the basis for the wind velocity maps shown in ASCE 7.

- *Fastest-mile wind*–the shortest time interval each day for passage of one mile of air flowing past the recording point.
- *Three-second gust wind*–the fastest wind-gust speed during any three-second period each day.

Wind speed typically is measured at a height of 33 feet in relatively open terrain. Data obtained at other heights are transformed to an equivalent 33-foot speed using the power law equation. Accordingly, all wind speed data from the U.S. Weather Service derive from velocities at 33 feet, i.e., V_{33}.

Recurrence Properties of Wind

Recurrence is a term which describes frequency of repeated phenomena. As applied to wind, it most often refers to the time frequency for velocities that are repeated or exceeded. One example, drawn from any commonly used wind map, is a mean (average) recurrence of 50 years for, say, a 100-mph wind, with a probability of non-exceedance of 0.02. This statement means that a wind velocity of 100 mph is expected to be exceeded 10 percent of the time every 50 years, on average (probability of exceedance = 50 x 0.02 = 0.1).

For buildings, the wind velocity of interest is the maximum value that reasonably can be expected at standard height in a particular location. Accordingly, attention is focused upon determination of an expected maximum wind speed at standard height during a selected time period. As already noted, that maximum velocity may not be the same from one location to another. Also, there is no way to know from historical records whether a worst-case wind velocity has occurred within the recording period, no matter how long that period might be. The concept of probability of recurrence provides the bridge across this problem.

Probability of Recurrence

Wind history provides a record of past maximum velocities, but it provides only a guide to the future. Probability methods are employed for projecting historical data into the future. The probability of recurrence is the basis for this determination. The probability of recurrence for various peak wind velocities in any chosen time interval can be established by statistical methods, using data drawn from past wind velocity records. The example of the 100-mph wind cited above is representative of such a probability estimate when projected into the future. This same probability of 0.02 mentioned above sometimes is stated in another way: There is a 90 percent probability that the velocity will not be exceeded in every 50-year period, on average. The 50-year period is called the mean recurrence interval. Likewise, a 25-year interval translates into a probability of non-exceedance of 0.04, and a 100-year interval translates into a probability of non-exceedance of 0.01.

Recurrence Interval

Statistical probability of recurrence requires that a time interval be selected for determining maximum expected wind velocity. Although any time interval can be used, wind velocity maps for building design purposes typically have used 25-, 50-, and 100-year intervals, with the 50-year interval being the most common. A representative, but assumed, 50-year life span for buildings provides one basis for this choice. The 50-year interval is called the recurrence interval.

Use of recurrence interval for establishing a wind

velocity to be used in building design presents several problems. First, most wind records eventually are broken as the time span of the historic record becomes longer. Therefore, the peak velocity predicted for any given time interval may not be the worst case in the future. Second, wind velocity fluctuates during short time intervals (a topic addressed next), and this data has not been extensively compiled, due to several factors such as past greater interest in sustained velocities, averaging techniques used in compiling the records and instrumentation sensitivity which has not always been capable of responding to gusts of short duration. Third, wind-recording stations typically are dispersed, so data for a specific building site may not be available.

These problems in using probability methods are not unique to wind applications. Indeed, determination of building live loads, snow loads, and earthquake-induced loads also have been determined from similar probability methods; they, too, introduce similar problems.

Moreover, a close reading of codes and standards will reveal that design loadings for all of these natural occurrences typically are based on a probability of 0.02. In fact, maps providing the load values usually refer either to the time interval or to a 0.02 probability of non-exceedance.

Notwithstanding the limitations, the recurrence interval is widely accepted and used as the best method available for establishing a wind-velocity standard for building design.

Wind Fluctuation

Discussion up to this point has proceeded with an assumption that the wind velocity is constant at a particular height regardless of the fact that the velocity is different at different heights of measurement. In fact, the wind velocity fluctuates continuously. It seldom blows at a constant rate, except at high altitudes where interference is minimal.

Wind fluctuation can be examined in several time frames:

- Short term, measured in seconds, called gusting.
- Cycles of several days, which encompass the time for passing of area-wide weather systems.
- Seasonal cycles, measured in months, caused by global weather influences.

From the standpoint of wind effects on buildings, pressures resulting from fluctuations over several days and seasons are those for which data has been historically gathered and mapped. Appropriate velocities can be selected through the procedures described above. Past methods of recording wind velocities have used averaging techniques to smooth out the velocity variations over longer time periods than seconds; e.g., the fastest-mile wind and the five-mile velocity (still another way wind data has been compiled). Gusting fluctuation data over time periods measured in seconds is not as readily available, if at all, from these records. Hence the historic records provide information about prevailing winds but limited information about gusting, although this situation is changing with the transition to three-second gust wind mapping.

Gusting

Gusting is a localized variation in wind velocity relative to the average velocity of the prevailing wind, and it generally means those velocities greater than the prevailing velocity. Gusting results from a variety of factors that include influences of both terrain and localized atmospheric variations.

Gusting is most obviously associated with storms, but even a steady wind will have gusting fluctuations in close proximity to the earth's surface. It should be noted that gusting in urban and suburban areas can be caused by building density upsetting the smooth flow of air.

For building applications, a major question on gusting to be answered is: At what point, if at all, does gusting in shorter time intervals become of greater concern in building loading than average wind velocities taken over longer time periods? The answer involves more than just adding the effect of a gust pressure to the building wind-loading. The methods of modeling and analysis of wind pressure to be used for building applications also are affected.

If gusting has no significant affect on structural behavior, then wind pressure may be treated as a static force. Static methods of analysis apply, and the design pressure is modified to reflect the gust effect. On the other hand, if gusting does affect building behavior in significant ways, then static methods may not accurately predict the building behavior, and dynamic methods may need to be used. It is to be noted that the concept of an equivalent static force is used under certain circumstances in seismic design and analysis.

Researchers of gust effects have suggested three generalized answers to the question posed above:

- When the gust wind is sufficiently large in area to encompass the whole building or building component. In this case, wind pressures on the various surfaces will cause an additional load on the entire building (or component).
- When the gust velocity causes a pressure on the building (or component) such as to affect the response of the building (or component). Depending upon the gust duration, this may result in dynamic loading of the building (or component).
- When the gust cycle frequency approximates or matches the natural frequency of vibration of the building or any of its higher order frequencies. The problem here is the possibility of resonance, whereby the gust-loading cycle amplifies the building response.

In general, a structure will not respond to the pressure caused by a momentary gust that encompasses only a small portion of the building. To affect structural response, a gust must have sufficient vertical and lateral extent to envelop the entire structure plus those air-stream patterns on all sides of the structure. A gust size with an extent of about eight times the building dimension in the direction of wind flow has been shown as necessary to transmit the associated pressures to all surfaces of the building. The lateral extent of a gust necessary to induce pressures on all sides of a building has been shown to be approximately two times the building width measured perpendicular to the direction of wind flow.

Wind Maps

The contours of wind velocity maps commonly found in codes and standards in past years typically have been based upon fastest-mile wind data. However, as mentioned earlier in the monograph, three-second gust velocity maps soon may serve as an accepted replacement. An understanding of the background on both types of maps in therefore necessary.

Since the velocity of the same wind varies in accordance with elevation above the ground surface, a standard basis of reference is needed for mapping. The standard basis is a height of 33 feet in relative smooth terrain, as is found at most airport locations. Wind velocities at other heights and in other terrain conditions then can be determined from tables or from the power law equation.

Maps based upon both the fastest-mile wind and three-second gust wind have several common aspects. Both provide velocity contours and wind velocities at a height of 33 feet; both are for open terrain; both maps are based upon an annual probability of 0.02 which the velocities shown will be equalled or exceeded.

To be noted, however, is that α exponents do not have the same values for procedures based upon the two mapped velocities (see Tables 2 and 3). Owing to the way the data has been compiled, exponents pertaining to the three-second gust map have been adjusted to fit empirical data associated with fastest-mile velocities.

WIND EFFECTS ON BUILDINGS

Wind Flow and Buildings

Deflection of air stream flow around obstructions was introduced in the previous section's discussion of wind pressure. That discussion was limited to a description of why pressure develops on surfaces as a consequence of stream flow interruption. In this section, stream flow and consequent pressures are examined in greater depth.

Two important concepts regarding wind flow follow from examination of the air stream. The first of these is the nature and distribution of pressures developed on building surfaces. The second is the concept of a pressure coefficient.

Effect of Shape of Obstruction

The flow of air around an obstruction may follow somewhat different paths, depending upon the shape of the obstruction. This effect is visualized by looking at the obstructions only in two dimensions. Resulting observations can be readily applied to objects of three dimensions in space, and this will be done when the obstruction shape takes on characteristics of building shape.

In Figure 9a, the obstruction shown circular in plan is a cylinder. The air stream is deflected by the circular shape, but the flow is relatively smooth. Wind velocity is increased slightly at the sides (represented by showing the filament lines more closely spaced). A curved obstruction produces a smoother air flow than one with angled faces.

The obstruction in Figure 9b is rectangular in

plan. In this case the air stream flow past the object is altered more severely. Additionally, the flow around sharp corners produces pockets where relatively still air seeks to merge with the altered air stream (shown in the diagram as curled flow lines at the locations of occurrence). Not only is the air stream deflected more in this case, its influence also extends further laterally. As well, the air stream on the side away from the direction of flow (hereafter referred to as the leeward side) reconverges but does so while drawing from a larger pocket of relatively still air. Lateral spread of the deflected air flow and downwind disturbances are increased as the width of the obstruction increases.

Turbulence

The garbled air flows at side and leeward surfaces are called turbulence and represent an erratic flow consisting of eddies of air attempting to mix with the prevailing flow. Mathematical descriptions of turbulent flow can be found in technical literature, but a precise quantification is not needed for the discussion here. The effect is seen in the diagrams.

Wind flowing past the circular obstruction is not without turbulence, but its degree is minor in comparison with the rectangular obstruction. The degree of turbulence increases as the curvature of the obstruction becomes more pronounced. It becomes even larger when angled surfaces replace curved forms and becomes still larger with obtuse angles. Lateral spread of flow interference also increases in the same way, and the leeward depth of reconvergence of the air stream is similarly affected.

Development of Pressure

Recalling that the air stream filament, which is represented graphically with an arrow in the direction of flow, also implies air velocity, it is to be noted that velocity necessarily changes in those regions where flow is interrupted. Pressures are developed on surfaces of the obstruction consequent to the varying velocities. Examination of these pressures reveals two characteristics of interest–their distribution and variation, and direction of the pressure vector, i.e., the directions of forces that are developed.

Pressure Distribution in Plan View

The plan shapes in Figure 9 are shown in Figure 10 with pressures rather than air stream flows. Characteristics of both pressure distribution and direction of the pressure vectors are observable allowing some generalizations to be made.

On the windward surface for both shapes, stagnation pressure level is approached, although less so for the circular shape. The pressure always acts normal to the surface, with the pressure vector acting toward the surface. In conventional terminology, this is called "positive pressure." The term "inward pressure" is sometimes used, but may be ambiguous in certain situations.

At the side surfaces, air velocity is greatest near the windward portions. This is much more pronounced for corner conditions. However, air velocity nearer the leading edges of side walls is less than the velocity further away–downwind along the walls. Along the sides, the pressure vector is in a direction

Figure 9a AIR FLOW AT CIRCULAR OBSTRUCTION

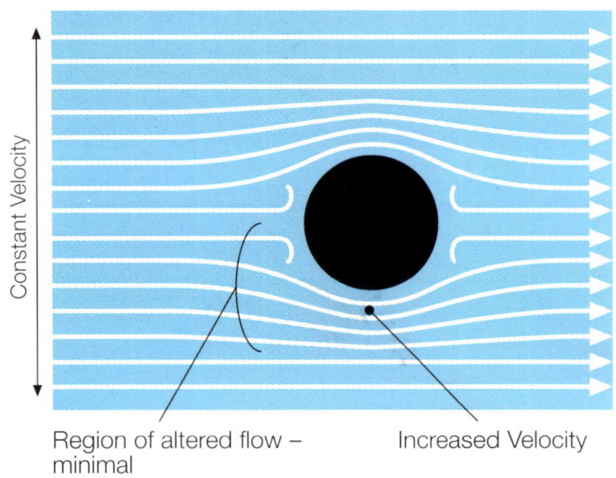

Figure 9b AIR FLOW AT RECTANGULAR OBSTRUCTION

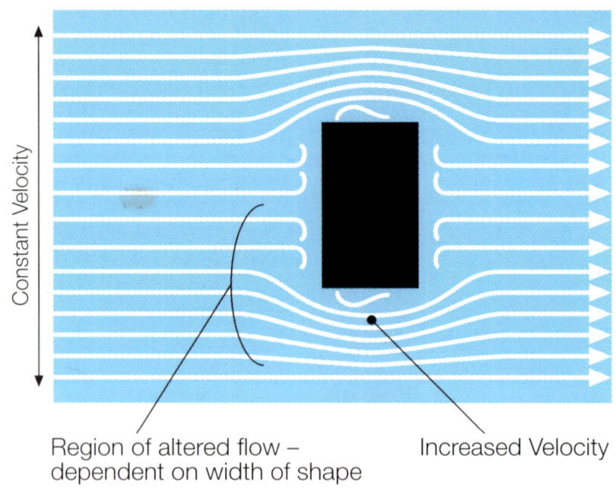

toward the region of higher velocity and acts perpendicular to, but in a direction away from, the surface. Such pressure acting away from a surface is called "negative pressure" (or "outward pressure" if used in conjunction with inward pressure, as previously mentioned). This characteristic is more apparent for the right-angle shape (Figure 10b) where the negative pressure commences at the windward corners. However, the same behavior may also be observed for the round shape (Figure 10a) where the curved surface is no longer exposed to the windward air stream.

Air movement on the leeward side and nearer the leeward surface is turbulent but eventually is pulled in the direction of prevailing air flow as the air stream reconverges. Here, again, the air velocity is lower nearer the corners, with negative pressures acting away from the surface.

The pressure variation on each surface varies in accordance with air velocity and shape and extent of the obstruction. While air pressure distribution and its vector directions are most readily seen for the right-angle shape, the same general behavior is observable for the circular obstruction's pressure profile. Obstructions of different shapes, proportions, and sizes have different pressure profiles, but overall behavior of air flow and pressures developed remain similar and are the basis for three significant generalizations:

- Pressure due to wind typically is developed on all surfaces of any obstruction placed in the path of flow.
- The distribution of pressure on any surface of the obstruction varies, and the variation depends upon size, shape, and proportion of the obstruction.
- The pressure vector acts normal to the surface and produces positive pressure on the windward surfaces of the obstruction and negative pressure on leeward surfaces and side surfaces.

Pressure Distribution in Cross Section

These same observations also apply to an obstruction viewed in a vertical section. Figure 11 shows vertical cross sections of a shape. However, in the diagrams of this figure the air flow is treated in the context of a building resting on the ground. Such a context produces somewhat different air flow behavior than occurs in plan view because of the ground surface, though the generalizations cited above still hold true. In particular, the surrounding ground surface causes deflected flow over the top of the obstruction.

Figure 10a PRESSURE DISTRIBUTION AT CIRCULAR OBSTRUCTION

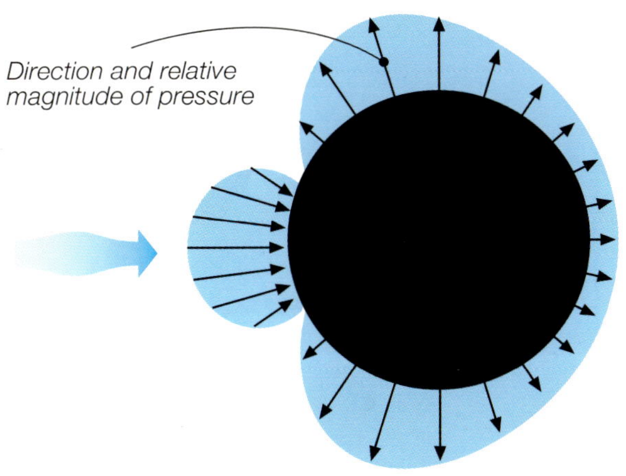

Figure 10b PRESSURE DISTRIBUTION AT RECTANGULAR OBSTRUCTION

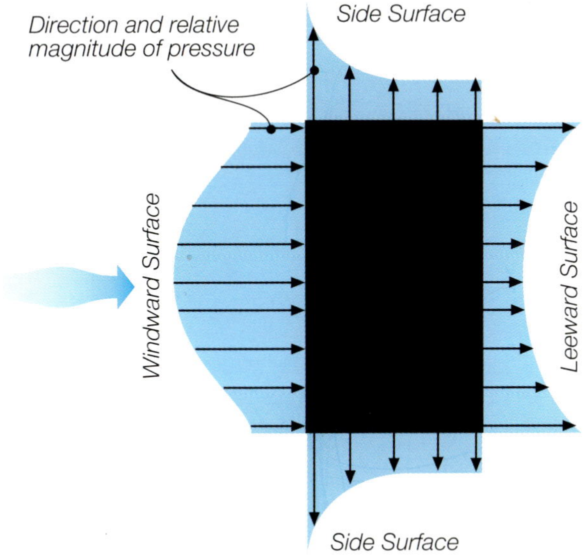

For a three-dimensional obstruction, the behavior discussed for plan views also occurs. On the windward and leeward surfaces, pressure vectors are in the directions as discussed above, but the pressure profile takes a different form due to both the boundary-layer effect and complete blockage of any air flow on the underside. The pressure profile for the top surface closely matches the side wall pressure profile for the rectangular shape discussed above, as it should.

Practical Treatment of Pressure Variation

Precise determination of pressures on surfaces obstructing the path of air flow is an extremely complex undertaking. Mathematical formulations for defining surface pressures are available for most simple geometric forms, but wind tunnel testing is necessary for most forms if accurate determination is needed. Because of the inconvenience and cost of wind tunnel testing, a simplified procedure for treating pressure variation is used for most, but not all, building applications.

Figure 11a AIR FLOW AT SIMPLE STRUCTURE

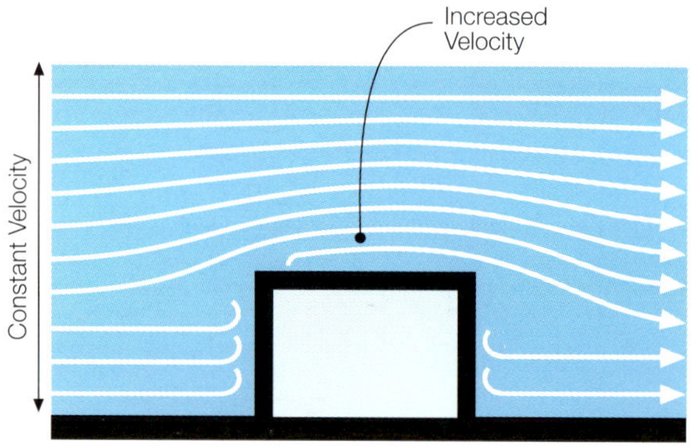

Figure 11b PRESSURE DISTRIBUTION AT SIMPLE STRUCTURE

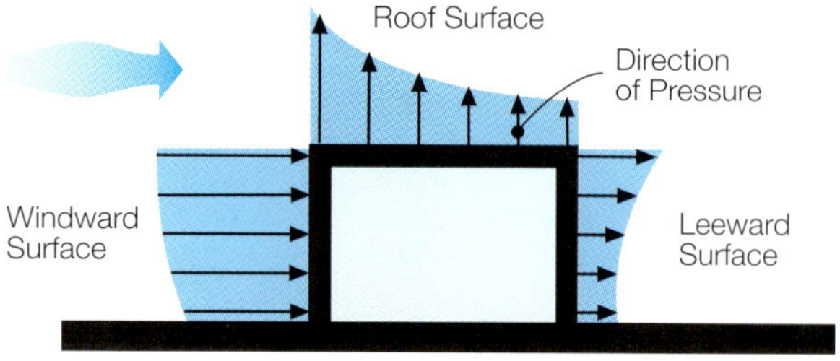

Pressure Coefficients

The simplified procedure referred to in the prior section utilizes pressure coefficients which provide an acceptably accurate modeling of the varying pressure. Typically, pressures calculated with the use of these coefficients are assumed to be static and apply to a static-force method of evaluation. Pressure coefficients are utilized for nearly all building applications except when engineering judgment calls for more accurate modeling of structural behavior. More precise modeling may be needed for building shapes where pressure coefficients are not available, taller buildings, or when there are other conditions involving dynamic behavior of the structure. Very tall buildings commonly are designed using wind tunnel test data rather than standard pressure coefficients.

Current standards set forth specific requirements for building proportions and height limitations for which the static-force method, which utilizes standard pressure coefficients, is to be replaced by an acceptable dynamic method. The height limitation typically is 160 feet with proportion limitations based upon an aspect ratio which is determined from building height and least plan dimension. In concept, the limiting proportion is that producing a fundamental building

Table 5 EXAMPLES OF PRESSURE COEFFICIENTS COMMONLY USED IN BUILDING APPLICATIONS*

SURFACE	PRESSURE COEFFICIENT
Windward Wall	+0.8
Leeward Wall	-0.5
Side Wall	-0.7
Roofs	
Flat (zero slope)	-0.7
30° slope–Windward Surface	+0.3 & -0.9
Leeward Surface	-0.7
45° slope–Windward Surface	+0.4
Leeward Surface	-0.7
60° slope–Windward Surface	+0.7
Leeward Surface	-0.7

* Coefficients are for main-frame wind loading in code applications.

frequency of one hertz or less, although current standards prescribe a specific limiting aspect ratio to simplify the work, such as 4:1 or 5:1.

The fundamental premise of a pressure coefficient is that an average pressure uniformly distributed over an entire surface, or a portion of it, can be used as a replacement for the varying pressure profile. By doing so, it is assumed that the effect of the average pressure reasonably approximates the effect of the varying actual pressure. This last condition is suitably satisfied for total wind loading on a building's surfaces. Use of an average value for the whole surface is less suitable for smaller, localized areas of the surface–for example, the windward corner of a side wall where the pressure is much larger than the pressures elsewhere along the wall.

The pressure coefficient (C_p) typically is a unitless value used as a multiplier of the wind pressure which applies; i.e., the design velocity pressure. The coefficient normally has a value of less than 1.0, but may be greater than 1.0 in some circumstances, such as when a localized area is considered.

Because the pressures and their vector directions usually are not the same on the several surfaces of a building, a separate pressure coefficient typically is used for each surface. Table 5 provides some typical pressure coefficients for the various surfaces of a rectangular-shaped building. Plus and minus signs are provided to indicate that the pressure acts toward or away from the surface.

Consider, for example, the closed building shape shown in plan and section in Figure 12. Typical pressure coefficients for all of the surfaces, as selected from the table, are shown at their appropriate locations. These apply to a building's primary wind-resisting system–the main-frame. The applicable wind pressure is multiplied by each pressure coefficient (from Figure 12) to obtain the pressure on each surface, shown in Figure 13.

Figure 13 shows the calculated pressures in their correct vector directions for q = 20 psf (the value is arbitrary and is for illustration only and does not yet reflect all considerations used in its determination). First to be noted is that all pressures are shown to be uniformly distributed on the surfaces; they are average values. Observe that the overall wind force acting on the building's primary wind-resisting structural system, often called the main frame, in the direction of flow is the sum of the forces on the windward and leeward surfaces. The diagram highlights this fact, even though the windward pressure acts toward its surface and the leeward pressure acts away from its surface. Note also that for this symmetrically shaped building the net force on the side walls is zero; that is, the total force on one side equals the total force on the other side but they act in opposing directions. In other words, there is tendency only for displacement of the building in the direction of air flow and upward, yet each surface is subjected to a unique average pressure. This use of average pressures explains why standard procedures distinguish between wind forces on a building's main-frame and forces on components. The average

Figure 12a PRESSURE COEFFICIENTS AT PLAN OF SIMPLE BUILDING

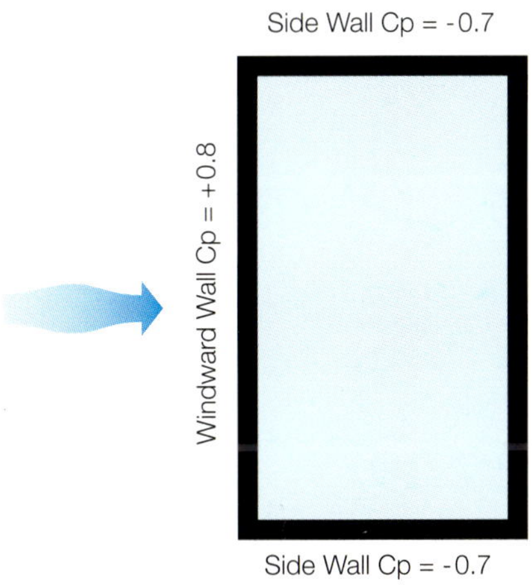

Figure 12b PRESSURE COEFFICIENTS AT SECTION OF SIMPLE BUILDING

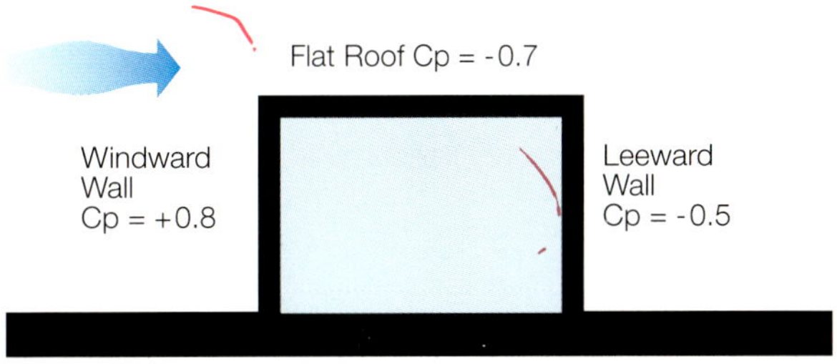

pressure is representative of the overall pressure but fails to acknowledge that certain portions of the building will experience pressures greater than the average value.

It is possible to select a pressure coefficient value which will provide the net overall pressure loading. In the example of Figure 12, an overall pressure coefficient of 1.3 (0.8 on the windward side plus -0.5 on the leeward side) multiplied by the same *q* value will yield the same net wind force as shown in Figure 13. Some references dealing with the general principles of pressure coefficients provide such overall values, but this is not standard procedure for building applications, primarily because of the need to develop wind resistance at all locations around a building.

Figure 13a CALCULATED PRESSURES AT PLAN OF SIMPLE BUILDING

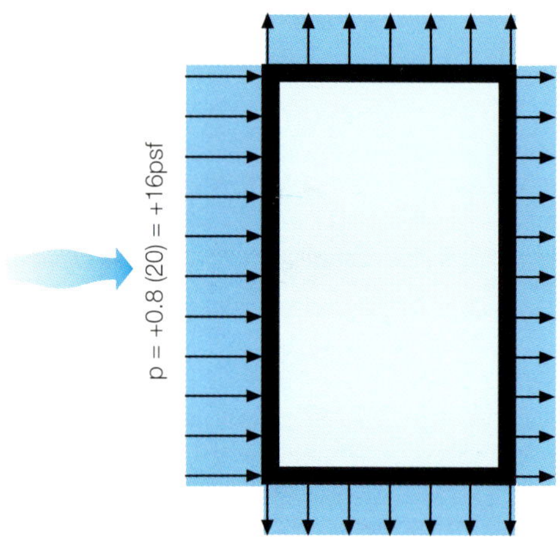

Figure 13b CALCULATED PRESSURES AT SECTION OF SIMPLE BUILDING

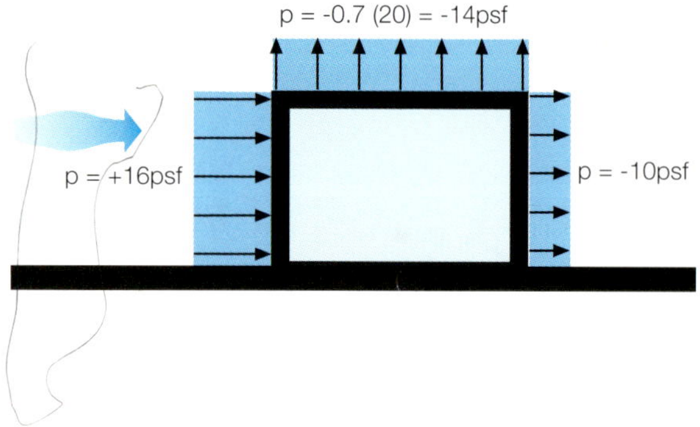

Pressure Coefficients for Roofs

In order to explain how pressure is developed and the non-uniformity of its distribution, examples to this point have illustrated only flat-top surfaces for obstructions. Building roofs have many profiles when viewed in cross section. Some roofs are flat, as in the previous examples, but many are sloped or may have various degrees of slope. It is useful to examine in greater detail the effects of sloped roof surfaces on air flow and consequent pressures. This is done in a general way in the diagrams of Figure 14, which show three different slopes–a relatively flat slope, a medium slope, and a relatively steep pitched roof. Air stream flows are indicated in Figure 14; the flow behavior is translated into pressure in Figure 15.

Clearly, the air stream flow is least altered from its steady state by the relatively flat slope. Indeed, if the slope is low enough, the developed pressures are not greatly different from the flat-roof case. There is a tendency on the windward slope for a positive pressure vector to develop, and this becomes more pronounced as the pitch increases. Also, on the leeward side increased turbulence caused at the peak tends to increase the negative pressure vector on that surface and becomes more pronounced with slope steepness.

Theoretically, there is a pitch at which the positive pressure vector on the windward surface exceeds the negative pressure vector. The negative pressure vector which tends to lift the roof surface (and also acts in a direction opposite to the air flow) changes to a positive pressure vector which tends to push downward and at the same time has a vector component in the direction of air flow. This condition is shown in Figure 15b for the roof of moderate slope. For the more steeply sloped roof, the changed condition is fully realized. The windward slope has a positive pressure and the leeward slope has a negative pressure. Figure 16 shows pressure coefficients for the same three roof slopes and pressure conditions of Figure 15. The windward side of the medium sloped roof shows two pressure coefficients–one each with a positive and negative sign, implying that pressures for both values are to be considered.

The theoretical slope at which the change from positive pressure to negative pressure occurs on the

Figure 14 AIR FLOW

Figure 14a LOW ROOF SLOPE

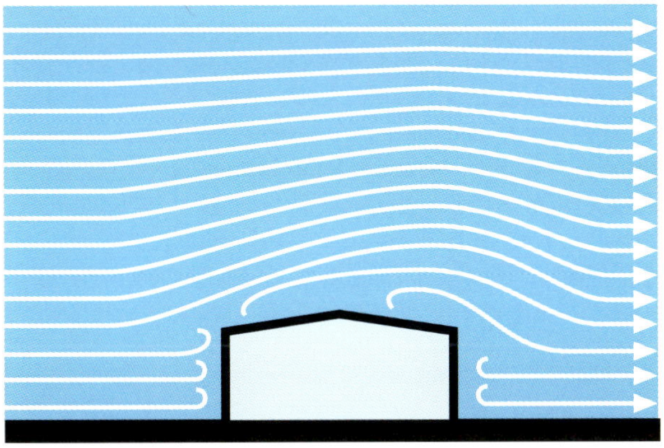

Figure 14b MODERATE ROOF SLOPE

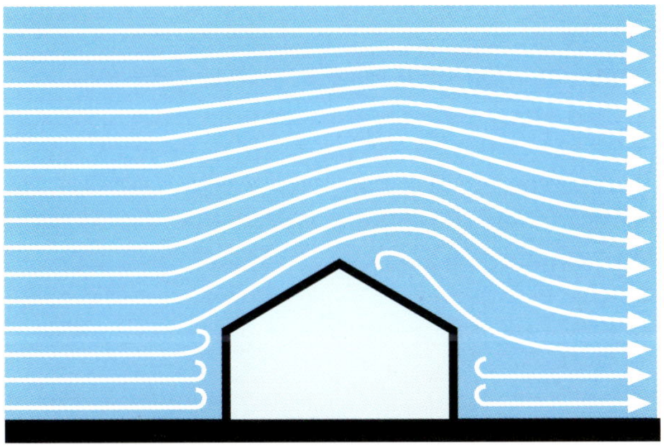

Figure 14c STEEP ROOF SLOPE

Figure 15 DEVELOPMENT OF PRESSURES

Figure 15a LOW ROOF SLOPE

Pressure profiles are approximate and will vary with roof slope and building dimensions

Figure 15b MODERATE ROOF SLOPE

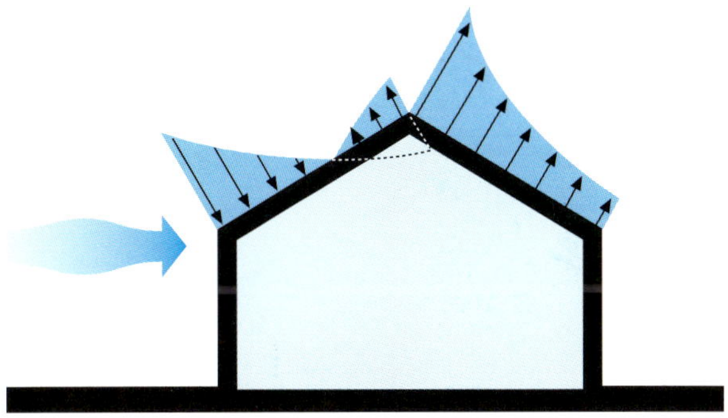

Figure 15c STEEP ROOF SLOPE

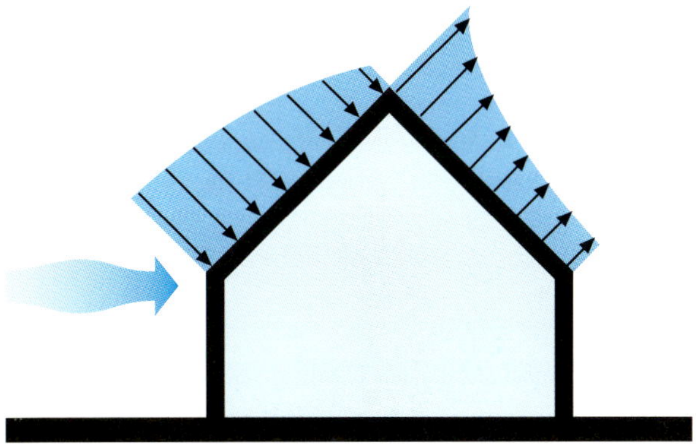

windward surface is difficult to determine precisely. Many conditions besides slope are involved, such as wind velocity, surface friction and size of the surface. For the selection of an appropriate pressure coefficient for sloped roof surfaces, this theoretical problem is answered in a general fashion by means of different coefficients which apply within specified ranges of slope. For the very flat slope (often described as having a slope of 10 degrees or less), a negative pressure coefficient is used.

Figure 16 PRESSURE COEFFICIENTS

Figure 16a LOW ROOF SLOPE

Pressure coefficents may vary in codes and standards depending upon slope and building proportions

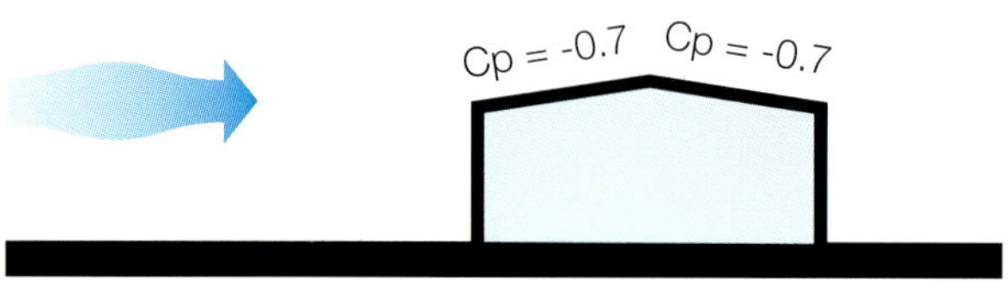

2:12 Roof Slope

Figure 16b MODERATE ROOF SLOPE

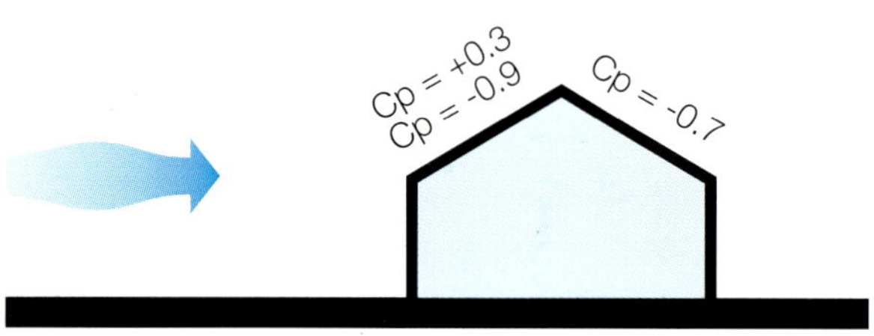

7:12 Roof Slope

Figure 16c STEEP ROOF SLOPE

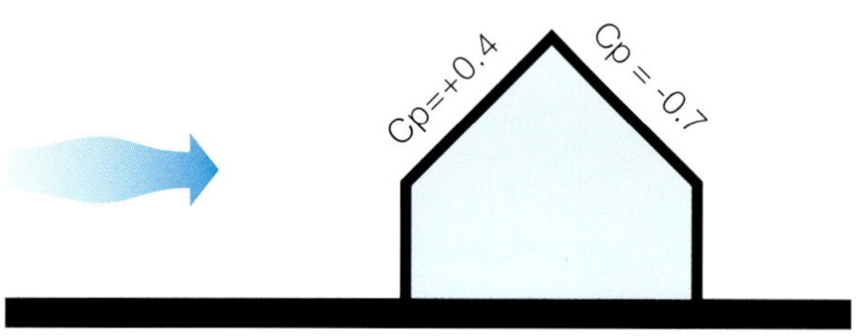

12:12 Roof Slope

For intermediate slopes, procedures in use call for consideration of two wind loading cases–first with a negative pressure coefficient, and then with a positive pressure coefficient. In other words, the forces developed from each condition must be acknowledged. For relatively steep slopes (usually described as at 45 degrees or greater), only a positive pressure coefficient is considered.

Pressure coefficients for varying roof slopes are not standardized in the various references. In the past, standard procedure has been to use three slope ranges and therefore only three sets of pressure coefficients. More recent design procedures in ASCE 7-95 expand both the number of different slope ranges and the associated values of the pressure coefficients.

Internal Pressures

To this point, we have discussed buildings as solid objects. However, this presents an incomplete picture in the study of wind forces since we have only discussed pressures applied to outside surfaces of buildings. But buildings also are in a sense hollow, with openings for access, ventilation and daylight. If buildings were fully closed and assured to remain fully closed, the discussion of fundamentals about pressure could be concluded at this point. Usually, buildings are not fully closed (windows and doors may not remain intact in some wind conditions), and the possibility and effects of pressures on inside surfaces–called internal pressures–must be treated.

If the free-standing wall shown earlier in Figure 4 were of finite dimensions, it should be apparent that wind pressures would develop on both windward and leeward surfaces of the wall. Consider again the rectangular shape from Figure 10b and the pressures developed on the windward and leeward surfaces. If the side wall lengths are reduced to just the thickness of the windward wall, the resulting pressures on the wall would be similar to those which will develop for the obstruction shown in Figure 11b without roof pressure, and the windward and leeward pressure coefficients would be approximately as shown in Figure 12.

If wind acts upon a building form of the same shape as that of Figure 12–except that a portion of the windward surface is omitted, as in Figure 17–the air stream passes through to the rear surface where it is blocked. The pressure on the inner side of the rear surface approaches stagnation pressure.

As the air seeks a flow path, pressures are developed on the inside faces of the windward and side walls, as well as on the inside surfaces of the roof and floor. The vector direction acts toward the surface in each case. These internal pressures on the facing wall, rear wall, side walls, floor and roof are additional to the external pressures shown in Figure 12. The net effect is determined by the pressure vector magnitudes and their direction on each surface.

Another illustration of internal pressure, but with different vector directions, is shown in Figure 18. In this case, a portion of the surface on the rear wall only (relative to the direction of air flow) is omitted, and the air flow pattern consequently is changed so that the tendency is to develop a suction inside the space. Air tends to be pulled out of the space at the rear whereas, in Figure 12, there was a wall surface. The directions of the internal pressure vectors act away from the inside faces of the windward wall, leeward wall, side walls, roof, and floor.

A third example of internal pressure is shown in Figure 19, in which a portion of one side wall is omitted. Once again a suction effect occurs as air tends to be pulled outward through the opening to merge with the air stream passing the building. These pressures also have vector directions acting away

Figure 17a AIR FLOW WITH OPENING IN WINDWARD WALL

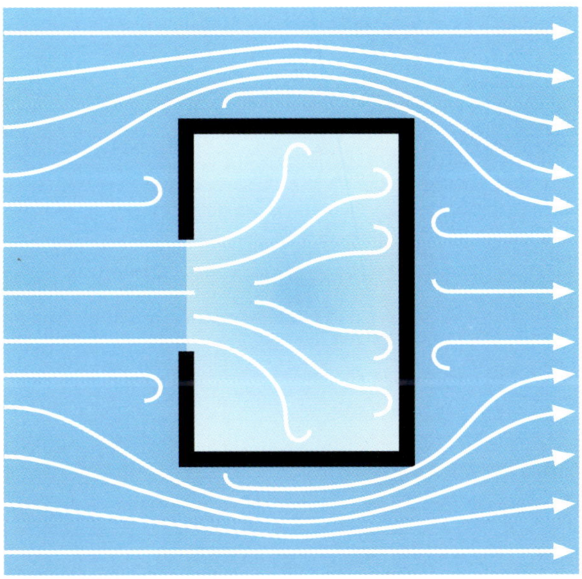

Plan

Section

Figure 17b INTERNAL PRESSURES WITH OPENING IN WINDWARD WALL

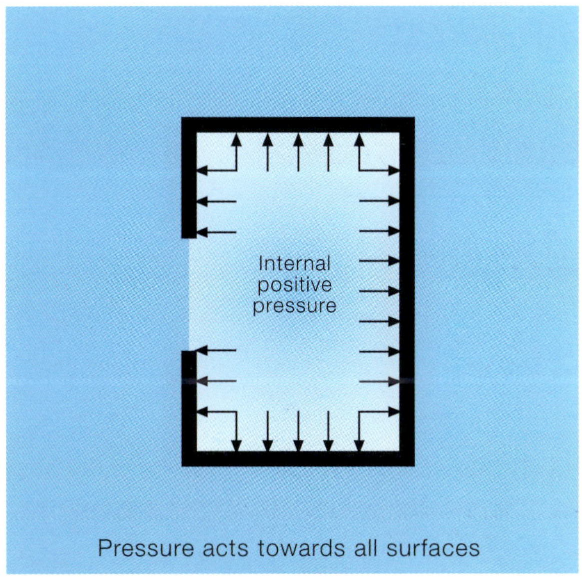

Plan

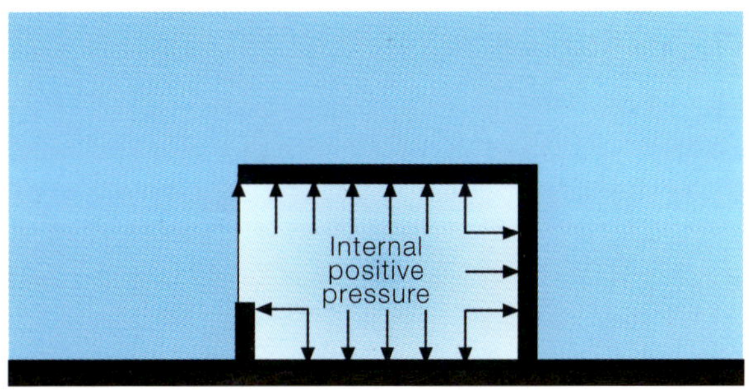

Section

from the internal surfaces as shown in Figure 18.

Two meaningful observations can be made from these examples. First, significant internal pressures will occur when surfaces have openings that permit through flow of air, and second, the directions of internal vectors depend upon where the openings are located relative to the direction of outside air flow; they have the potential to produce forces that may tend either to explode the building or to cause its inward collapse.

Internal Pressures and Buildings

Most enclosed buildings have windows and doors in exterior walls. If these openings remain securely closed, internal wind pressures will not develop, and in most cases, there would be no need to give them further consideration.

However, when extreme winds occur, failure of openings–especially glazed areas–is a distinct possibility and often happens.

Ordinary window glass cannot be counted on to remain intact, particularly during hurricane winds, and wind pressure alone on glass is not the only reason. Extreme winds produce a considerable amount of airborne debris which routinely causes glass breakage.

Given the possibility that windows and doors may

Figure 18a AIR FLOW WITH OPENING IN LEEWARD WALL

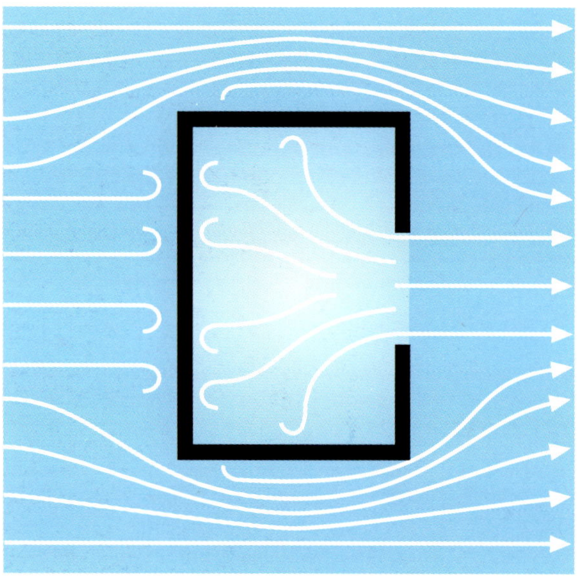

Plan

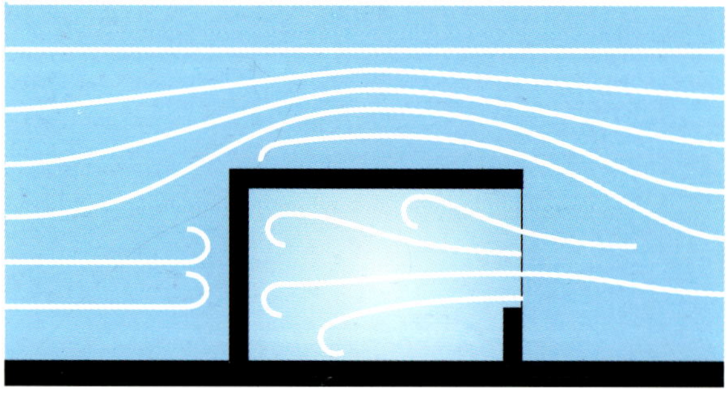

Section

Figure 18b INTERNAL PRESSURES WITH OPENING IN LEEWARD WALL

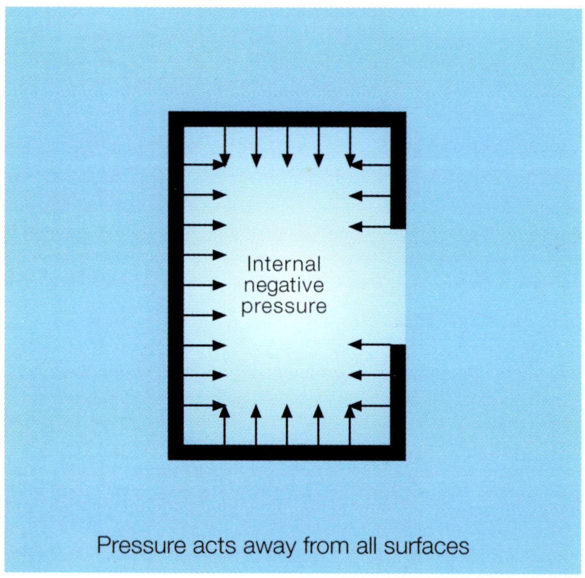

Plan

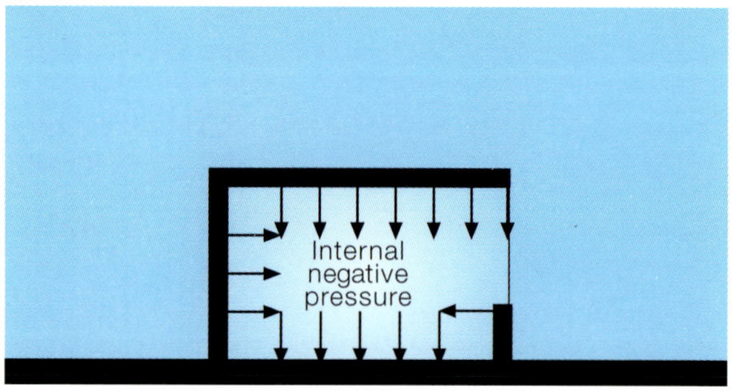

Section

fail in extreme winds, good design judgment requires architects to consider internal pressures as well as external pressures. Failure to consider the internal pressures invites major building damage when high-velocity winds occur. This is not new information, and standard references have long specified their consideration. Too often, these internal pressures are ignored by architects and engineers. Owing to the extensive wind damage in recent hurricanes, especially Hurricane Andrew, more rigorous requirements are being introduced to ensure that the possibility of internal pressures are considered. Standards have been developed regarding strength as well as puncture resistance for windows, cladding and other potentially vulnerable enclosure materials. In hurricane-prone counties of south Florida, designers are now required to take internal pressures into account when these standards cannot be met.

In the past, buildings have typically been considered as either open or closed. This definition has been revised because of the uncertainty of what is meant by an opening in a building. Further definition is provided in more recent standards, and no building is considered to be completely closed. Openings of two types are generally accepted–permanent openings and glazed openings, and internal pressure coefficients are applied to each case.

Permanent openings in exterior walls or roofs allow wind to flow into the building, such as air intakes or exhausts, gaps around doors, gaps in cladding and certain glazed openings as differentiated in the next category.

Glazed openings are closed by glass or plastic, such as windows (fixed or operable), skylights and glass doors, which in regions of extreme wind velocities, either must be designed to withstand pressures and missile loads (as defined in references) or otherwise must be considered as permanent openings.

We have yet to address internal pressure magnitudes or the effect of size and distribution of openings on internal pressure. In practice, coefficients also are used for these pressures. Sizes and distribution of

Figure 19 AIR FLOW WITH OPENING IN SIDE WALL

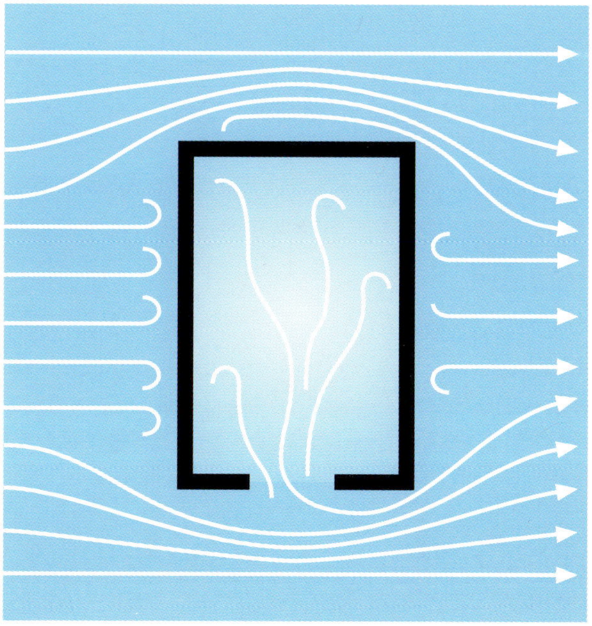

Plan

Photo 4 ***Debris in extreme winds such as hurricanes frequently punctures cladding and doors, as seen above.***

Figure 20 ZONE OF INCREASED PRESSURE COEFFICIENT AT CORNER CONDITION

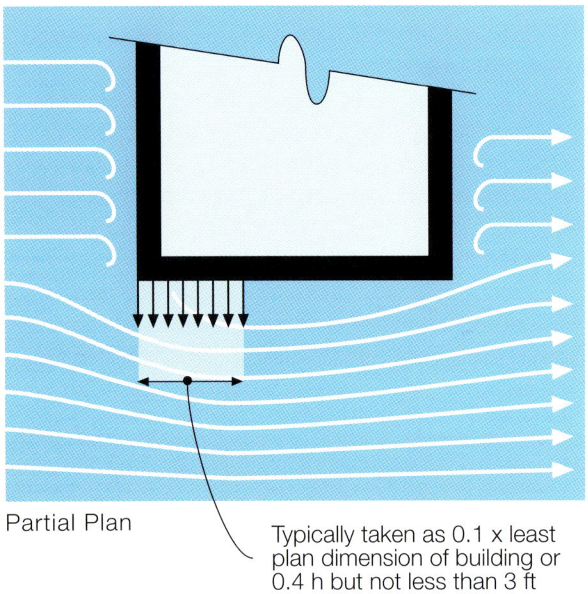

openings in perimeter walls are a consideration in selecting appropriate coefficient values from tables, including the directions of the pressure vectors. In application, internal pressures contributing to overall loading of a building are added (considering vector direction) to external pressures because the two are determined by slightly different procedures.

Until now, our discussion has focused on wind pressures developed on overall surfaces of buildings. These pressures pertain to two aspects of a building's resistance: the capability of the building as a whole to resist those forces which are developed, including individual building elements that participate in carrying overall loads–typically called the main wind-frame and the overall strength of entire surfaces and their participating load-carrying parts; i.e., walls and roofs. Our discussion so far has not addressed complete design pressures–that is, the actual pressure loadings used for building design purposes. These design pressures are covered in the discussion on application procedures.

Wind Pressures on Building Components

The general discussion of wind pressures developed on the various surfaces of an obstruction (Figure 10) pointed out that the actual pressures are not uniformly distributed as assumed with use of pressure coefficients.

Figure 21 ZONES OF INCREASED PRESSURE COEFFICIENT AT ROOF CONDITION

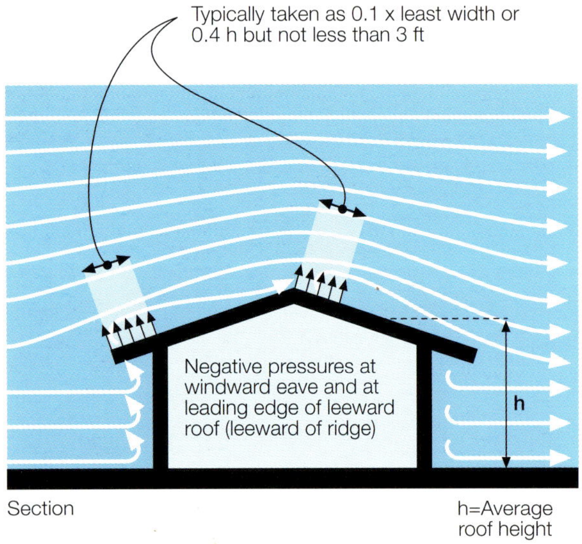

Figure 22 UPLIFT PRESSURE AT ROOF OVERHANG

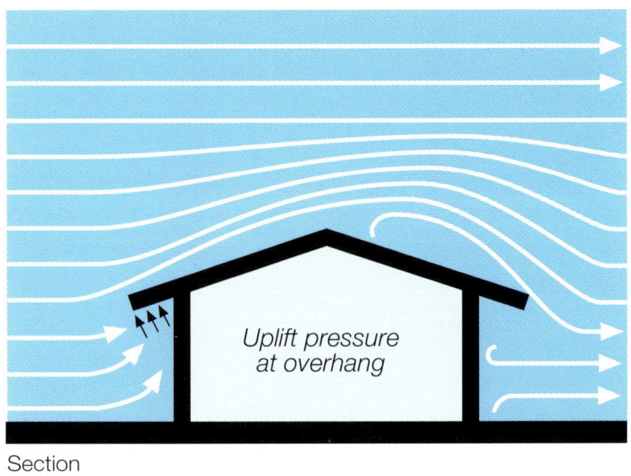

There are some regions where pressures are much greater than an assumed average pressure; these regions are dealt with separately. The usual terminology, when discussing these localized regions of increased pressure, is "components and cladding." This is a distinctly different use of the term "component" as applied in the context of a building's main-frame.

Building Corners

Figure 10 also shows that the side wall pressure is actually largest at the leading corner. Use of pressure coefficients uniformly over an entire surface fails to consider those pressures that exceed the average value. Frequently, a portion of the side wall length, taken at the windward corner, may have an average pressure exceeding by several times the average pressure for the entire wall. Figure 20 shows at larger scale this corner condition and its associated pressure. Since building proportions have an effect upon the magnitude of the pressure increase at the corner, the actual amount of increase varies for different conditions.

Roof Edges

A similar effect at the leading edge of a roof occurs, as is illustrated in Figure 21. If the roof has a gabled form (as in this figure), an increased pressure also develops on the leading edge of the leeward roof surface, the ridge.

Roof Overhangs

Building configurations rarely match the simplified diagrams used in describing fundamentals of wind behavior and pressure development. External appendages and construction features often alter the air stream flow from the basic cases used in examples. A roof overhang is a particular kind of variation for which wind effects must be considered. Although many roof configurations may be designed, only the simple roof overhang is treated here. This configuration is highlighted because of its prevalence and because the consequent pressures affect a particularly vulnerable part of a building.

A typical roof overhang condition is represented diagrammatically in Figure 22. The overhang on the windward side is the focus of consideration. Figure 22 shows that air flow around the windward wall includes an upward moving stream which is partially blocked by the overhang, thus producing an upward directed pressure on the underside of the overhang. This pressure effectively is the stagnation pressure of the wind, although the actual magnitude depends upon the depth of the overhang. On the upper side of the roof surface there also is the external roof pressure which extends across the overhang.

The net effect of these two pressures is the vector sum of the two pressure loadings acting on just the overhanging portion. Since the vector direction of the pressure on top of the overhang may be upward or downward, depending upon roof slope, the net effect must include the pressure vector directions.

Localized and Progressive Failures

These examples of increased pressures provide the reason for localized failures commonly associated with winds of high velocity and are illustrated in photographs included in this monograph. Plywood torn from roofs can typically be traced to leading edges of surfaces relative to the wind direction, as can siding torn from corners. Photo 5 provides a graphic illustration of localized tearing due to pressure increases at roof edges.

Localized failures often lead to more extensive wind damage. Greater damage results when other building components are exposed to an altered air

Photo 5 ***Composition shingle and underlayment failure.***

stream flow brought about by the initial failure. This sort of occurrence is called progressive failure and usually leads to more severe damage; in the worst case, it can lead to total building collapse. Localized pressure increases are quite large and occur on virtually every building. They cannot be ignored.

Main Wind-Frame and Components/Cladding are Separately Treated

Practical treatment of localized pressures for individual building components is simplified from pressure vector summations, as described for roof overhangs. Accepted procedures separate wind design considerations into two basic parts: overall wind loading of the main-frame and separate consideration of wind effects on particular components and cladding. With this approach, pressures affecting the components and cladding are not additive to those applied to the main-frame.

The overall forces are determined in accordance with theory covered in the preceding sections. Localized pressures are determined using wind velocity data as described earlier, but using pressure coefficients associated with specific components and locations. Pressures determined for localized conditions are not additive to overall pressures, because the two design purposes are different. Design for overall wind loading assures adequate strength and performance for the building as a whole. Proper design for localized pressures will assure performance of individual components of a building.

Of the two considerations–overall resistance and component resistance–architects usually have a greater responsibility for the latter, although overall building strength is the more important life-safety concern. Engineering consultants are usually employed to look after the overall wind resistance for a particular building either to do the necessary structural design completely or, for simpler buildings, to provide technical guidance to the architect.

For the problem of localized pressures, however, strength concerns are associated as much with connections in building assemblies as with the structural capacity of members of the assemblies. How often does an architect consult with a structural engineer in the course of detailing roof eaves, cladding, or window assemblies? For most architects, rarely, with the possible exception of large commercial and institutional buildings. Yet the problem remains that connections and assemblies are the source of localized failures.

Since the solution to this problem is directly linked to everyday architectural practice, this monograph includes a section on building components that discusses detailing and construction assembly considerations. The monograph concludes with a listing of recommended practices that address wind resistance by building assemblies.

PROCEDURES FOR WIND-RESISTANT DESIGN IN PRACTICE

The preceding discussions have examined fundamental characteristics of wind acting upon buildings. For the most part, these fundamental characteristics are theoretically based, although diagrams were used in lieu of mathematical formulas. Through the use of mathematical formulations of these concepts, it is possible to analyze a building for wind effects, but this is not the way it is normally done in practice. Instead, standardized procedures and tables of values for design pressures and coefficients are used. In the discussion which follows, fundamental concepts are placed into procedural context more common to architectural practice.

There are several standard procedures in use throughout the United States; each of the three model building codes and ASCE prescribes one, and all are different. Procedures in current editions of the BOCA and SBCCI (Standard) Codes resemble the procedures of ASCE 7-93, but procedures of the ICBO (UBC) are significantly different. ASCE 7-95 retains many procedures of ASCE 7-93, but adds new procedures for specific situations and changes all table data to correlate with three-second gust winds. The procedure to use often is dictated by the applicable building code. Although fundamental characteristics are consistent among these procedures, the manner in which they include and combine data often are not. In this monograph, we face a dilemma that any discussion of procedures of application must treat these issues of difference in the several standards. Yet, by stopping short of a discussion of all applications procedures, topics meaningful to everyday architectural practice are missed.

The jump from theory to practice is further complicated by the way in which the foregoing concepts are, in fact, treated in application procedures. Efforts to simplify the material for easy, everyday use have resulted in tables of values which are provided in all codes and standards.

A problem with tabled values in place of fundamentals is that there is inconsistency both in symbols and in considerations included in tabled values. For example, the distinction regarding internal pressure development is obscure at best in standard procedures. The internal pressure effect typically is taken into consideration, however, and appears in coefficient values, as will be pointed out in the next sections. Tabled velocity values also are presented in different forms in the codes and standards.

Separate treatment of each of the codes and standards is, therefore, not attempted in this monograph. Instead, procedural concepts are covered that are usually, although not always, common to the several standards. When particular concepts are treated using symbols as a means for abbreviating the discussion, readers must realize that symbols are different in the several reference standards. Moreover, corresponding symbols in the references sometimes include or omit different parts of the concepts. When symbols are needed, either for clarity or brevity, those from ASCE-7 are used in this monograph. When data from tables are needed in a particular discussion, these also are from ASCE-7.

General Considerations of Standard Procedures for Wind Force Analysis

Standard procedures for building design to resist wind are much the same in their overall methodology. Most of the differences which occur are in the assembly of wind data, additions or omissions of prescriptive requirements and degrees of rigor applied in the determination of some wind effects. These common considerations in methodologies are first identified; then each is discussed in greater detail.

General considerations for overall wind effects and their effect on components are treated separately to avoid any confusion that might arise from combined treatment, as often is done in the codes and standards. Considerations in design for wind effects on buildings can be viewed as consisting of the following steps:

1. Determination of a *basic wind pressure–q:*
   - Selection of an appropriate *basic wind speed–V.*
   - Selection of an appropriate *exposure category–A, B, C, or D*
   - Selection or calculation of *q*–pressure at standard height of 33 feet.
2. Determination of *velocity pressures*–also called *q* and usually distinguished with subscripts:
   - Selection of an appropriate category of *building importance–I.*
   - Selection or calculation of *velocity pressure(s)*–also called *q* but with consideration of height, *I*, and gusting (procedures vary at this point, and several velocity pressures may be needed.)
3. Determination of *design wind pressures–p:*
   - Calculation of design wind pressures for the *main wind-force resisting system*–including positive and negative pressure effects on all exterior surfaces.
   - Calculation of design wind pressures for building *components and cladding*–including positive and negative pressure effects.

Step 1 leads to the determination of a basic wind pressure which applies to the particular building based upon the wind environment at the site. This basic pressure then is used to complete Step 2, which takes into account use or occupancy of the building and velocity variation with building height. The design pressure for each surface is determined (Step 3), taking into consideration the location of the surface relative to direction of wind flow *(pressure coefficient C)* and gusting effects on the surface *(gust factor G).*

Basic Wind Speed

The wind speed used for a building's wind-force evaluation is usually, but not always, obtained from a map included in the code or standard that is used. Velocity values are given for winds that are horizontal to the earth's surface. Maps may vary from one reference to another and are updated periodically. Velocity contours typically indicate extreme winds for a 50-year recurrence of 90 percent probability of non-exceedance.

Wind speed maps typically show the entire United States on a single page and do not distinguish small areas where local extreme winds may differ from the "big picture." Some maps may identify areas where larger wind velocities require special consideration; however, the local building authority is usually the source of wind speed information in such cases. Additionally, local and state building authorities may have wind speed requirements, which will usually prevail over the maps provided in standards.

The basic wind speed is modified to reflect terrain conditions at the building location. It is further modified (Step 2) for the surface height, and more than one height may be necessary (this will be explained as the discussion progresses).

This modification is accomplished in different ways and is based on procedures in the particular code or standard. Velocity pressure is then obtained for each applicable value of *q*. Procedures for this also are different, depending upon the code or standard being followed.

Exposure Category

The exposure category describes terrain roughness conditions surrounding the building site and four exposures identified earlier are now commonly accepted. Their designations and descriptions are as follows.

Exposure A — Large city centers with at least 50 percent of the buildings having a height in excess of 70 feet, and when these conditions prevail in the upwind direction for a distance of at least one-half mile.

Exposure B — Urban and suburban areas, wooded areas, or other terrain with numerous closely spaced obstructions having the size of single-family dwellings or larger, and when these conditions prevail in the upwind direction for a distance of at least 1,500 feet.

Exposure C — Open terrain, including grasslands and flat open country, with scattered obstructions having heights generally less than 30 feet.

Exposure D — Flat unobstructed areas exposed to wind flowing over open water for a distance of at least one mile and extending inland from the shoreline a distance of 1,500 feet, and applies when the building is exposed to wind coming from over the water.

Not all of the exposure categories identified above appear in each reference; for example, Exposure Categories A and D sometimes are omitted. This simply reflects differences or ongoing changes in the codes and standards. However, none include more than the four exposure categories listed.

All of the codes and standards use Exposure Category C (open terrain) for defining the basic wind speed, but this may not be apparent in some references, due to the manner in which the wind data is presented.

Exposure Category C, of course, is also the basis for wind speed records.

Building Importance

This topic is a part of all codes and standards and a design factor. The concept of building importance provides for consideration of buildings regarding their functional purposes and the importance society places upon these functional purposes. Some buildings may have higher life-safety risks, due to type (duties, infirmities, etc.) or numbers of occupants, or they may have critical social purposes, such as health-care facilities, fire stations, and community emergency functions–as contrasted with buildings of low occupancy and day-to-day residential or commercial use. For buildings deemed especially important for one or more of these reasons, a coefficient is introduced which, in effect, results in larger design forces, and thereby ensures that such buildings will have greater resistance to wind. Similarly, for buildings having little or no human occupancy, wind design forces may be lower.

Typically called an importance factor, and represented with the symbol *I*, the coefficient is attached as a multiplier when determining the velocity pressure for a building surface. The effect of this coefficient is to increase or reduce the design pressure. Larger pressures result for buildings of greater importance, and design pressures for buildings that represent a low hazard to human life may be lower than normally computed.

The importance factor is attached to the velocity-pressure equation in different ways in the various references, either as a linear coefficient which is a direct multiplier, or as a coefficient attached to the velocity whereby the basic pressure value *q* is modified before the velocity is squared, or as a part of the velocity whereby the velocity is modified before squaring. Accordingly, there are no standard values for the Importance Factor coefficient because they are determined by how they are used in the pressure equation. Typically, four levels of building importance are used.

These four levels of importance are much the same in the referenced codes and standards, though their coefficient values are not. These levels, designated by Roman numerals in Table 6-2, ASCE 7-95, are described as follows:

- I Buildings that represent a low hazard to life in the event of a failure, such as agricultural buildings, temporary buildings, and minor storage buildings. This coefficient usually has a value of slightly less than 1.0.
- II Residential, smaller commercial buildings and similar buildings of ordinary everyday use. This is considered the normal case that applies to most buildings. This coefficient will typically have a value of 1.0.
- III Buildings having a primary occupancy in which more than 300 people congregate in one area. This coefficient will typically have a value of slightly larger than 1.0.
- IV Buildings designated as essential facilities, including hospitals, fire and police stations, communication centers, power stations, and emergency shelters. This coefficient will typically have a value of greater than 1.0 and is the most restrictive case–with the largest design wind pressures.

Note that use of an importance factor (except for Case II) will alter the velocity pressure and design wind pressure even though the basic wind speed remains the same. Also note that in wind design, values of I for application to wind are not the same as values for a similar concept used in earthquake-resistant design.

Wind Speed-up Over Hills and Escarpments

Wind has a tendency to increase in velocity as it flows up a steep hill and passes over its crest. This characteristic, although long known, has not been covered in wind code procedures until recently. It is now treated in the most recent standard. In ASCE 7-95, based on three-second gust winds, consideration is given to localized increases in wind velocity caused by small hills and ground escarpments. The increased velocity is negligible and may be ignored for slope ratios less than 0.2, or when the height of the rise is less than 15 ft (based on the hill height relative to a particular upwind horizontal dimension and the exposure category); but otherwise, it is considered for analysis based on ASCE 7-95 procedures.

When the increased velocity is considered in an analysis, there are three independent multiplying factors to take into account. One is for the topographic effect, where the main considerations are the type of hill and the slope of its rise. The second is the distance from the crest of the hill to the building. The third accounts for the height of the crest of the hill above the surrounding local terrain. These three factors are multiplied together and then added to 1.0 to produce a topographic factor K_{zt}, which is attached as a multiplier to the velocity pressure q (the three-second gust velocity). This procedure produces an increased velocity that is based on the three considerations noted and from which velocity pressures are calculated.

Wind Pressures for the Main Wind-Force Resisting System

We see now that the design wind pressure involves consideration of wind velocity at the applicable height(s), terrain conditions (building exposure), the building importance factor, the applicable pressure coefficient and gusting. All of these considerations are combined into an equation from which the design wind pressure is calculated. Depending upon the code or standard, these considerations may or may not be unique factors in the basic pressure equation. Indeed, this is where the principal differences among the references are found and causes difficulty in providing a step-by-step procedure that is applicable to all of them. Even so, a practical understanding of wind forces on buildings requires a general understanding of how these considerations are treated.

Overall Pressures

Overall pressures are for the purpose of analyzing the building's primary wind-resisting system–the main frame. Overall pressures are used not only for design of the main frame (entire building structural system), but also for portions of, or entire individual surfaces. ASCE 7-93 treats pressures on whole surfaces; whereas ASCE 7-95 subdivides the surfaces. Regardless of how they are treated, these pressures are the ones used in design of the main wind-force resisting system.

A general formula for determination of the basic wind design pressure for building surfaces is as follows:

design pressure = velocity pressure x pressure coefficient x gust factor

$$(p) = (q) \quad (C) \quad (G)$$

The velocity pressure q is the pressure for the applicable exposure condition and height, and which sometimes is adjusted to include the building importance factor. The term "velocity pressure" is used, rather than "stagnation pressure," since additional considerations are included when arriving at the value. C in the equation is the appropriate pressure coefficient, already discussed. The gust factor G, which in some procedures is combined with C, is further discussed in the monograph.

In the equation above, the velocity pressure q in ASCE 7-95 is defined as follows:

$q = 0.00256\ I\ V^2$

Note that the importance factor is attached to the equation for stagnation pressure which was provided earlier, and it is not attached to the velocity before squaring as in ASCE 7-93.

Gust Response Factor

The symbol G appearing in the design pressure equation above takes into account wind velocity increase as a consequence of the gusting behavior of winds. Height above the ground surface is the principal practical consideration in determining what coefficient value will be used. Also, coefficients are different for procedures based on fastest-mile wind and three-second gust winds.

Complex mathematical expressions have been developed to quantify the gust effect on buildings. No further discussion of these is provided in this monograph; however, if greater detail is needed, it can be readily found in pertinent technical literature, including ASCE 7-93 and ASCE 7-95.

Of more immediate interest and useful to the purposes of this monograph is how gust considerations are actually introduced into standard procedures for determining wind pressures.

The effect of gusting on buildings appears in standard static procedures for wind analysis in the form of the Gust Response Factor, G. This factor accounts for additional loading due to above-average wind velocity. It also takes into account, though in simplified form, the randomness of gust loading and the fact that gusting is also influenced by ground surface conditions (height) as well as dynamic response of the structure. The factor typically appears as a multiplier coefficient in a wind pressure equation.

Sometimes the factor is obtained from tables; sometimes it is in plotted graphs; sometimes the additional gust loading effect already is included with other constants in tables used for determining the design pressures.

Values of G may be provided in tables which consider both height and the exposure category.

Windward Wall External Pressure

Because many terms in the design pressure equations are uniquely determined when actually applied to buildings, their symbols are subscripted in order to clarify the consideration making them unique. In subscripted form, and in accordance with the particular current standard indicated earlier, the equation forms for a windward wall surface are:

$p_z = q_z C_p G$

$q_z = 0.00256\ I\ V_z^2$

The subscript z clarifies that the pressure is at a particular height above the ground surface and so recognizes that the pressure varies with height. Its attachment to the stagnation pressure provides similar acknowledgement that the velocity pressure is also for a particular height.

The subscript p in C_p clarifies that the pressure coefficient is for external pressure developed on the windward side of the surface. Algebraic signs are applied to the pressure coefficient in accordance with the rule stated earlier in the monograph–namely, that pressure acting toward the surface is positive (+) and when acting away from the surface is negative (-). On the windward exterior surface, the pressure acts toward the surface and therefore has a positive sign.

Values for the pressure coefficient C_p and the gust effect factor G normally are provided in the code or standard being used, and for some applications they are combined into a single term (GC_p). The applicable pressure coefficient is that on the windward surface.

Wind velocity is considered for the applicable exposure condition. Because of the way data are provided in some references, the equation for velocity pressure sometimes appears as:

$q_z = 0.00256\ K_z\ I\ V^2$

In this form, wind velocity values V are for Exposure C (open terrain), and the term K_z modifies the velocity to the applicable exposure and height condition. Both forms yield similar velocity pressures, provided that the correct velocity value is used.

In ASCE 7-95, this same expression includes the topographic factor mentioned earlier and takes the form:

$q_z = 0.00256\ K_z\ K_{zt}\ I\ V^2$

Velocity pressure equations for leeward and side walls are discussed later and examine an important simplification in practice.

Stepped-Pressure Profile

The equation for velocity pressure given above provides wind pressure values for the windward wall at every height simply by using that height in the calculation. However, obtaining the pressure profile in this way, while perfectly valid, will require a number of tedious calculations. Moreover, the resulting pressure profile cannot be readily transformed to forces, due to the complex mathematical form of the profile curve. Practical applications proceed in an easier way through simple assumptions. The simplification is use of a stepped-pressure profile which has constant pressures for specified height increments.

Height increments for the stepped-pressure profile typically are specified in the code or standard being used. Normally, ranges of above-ground heights are specified–for example, from zero to 15 feet, greater than 15 feet but not greater than 20 feet, and so on.

Figures 23a, 23b, 23c STEPPED PRESSURE PROFILES–WINDWARD WALL

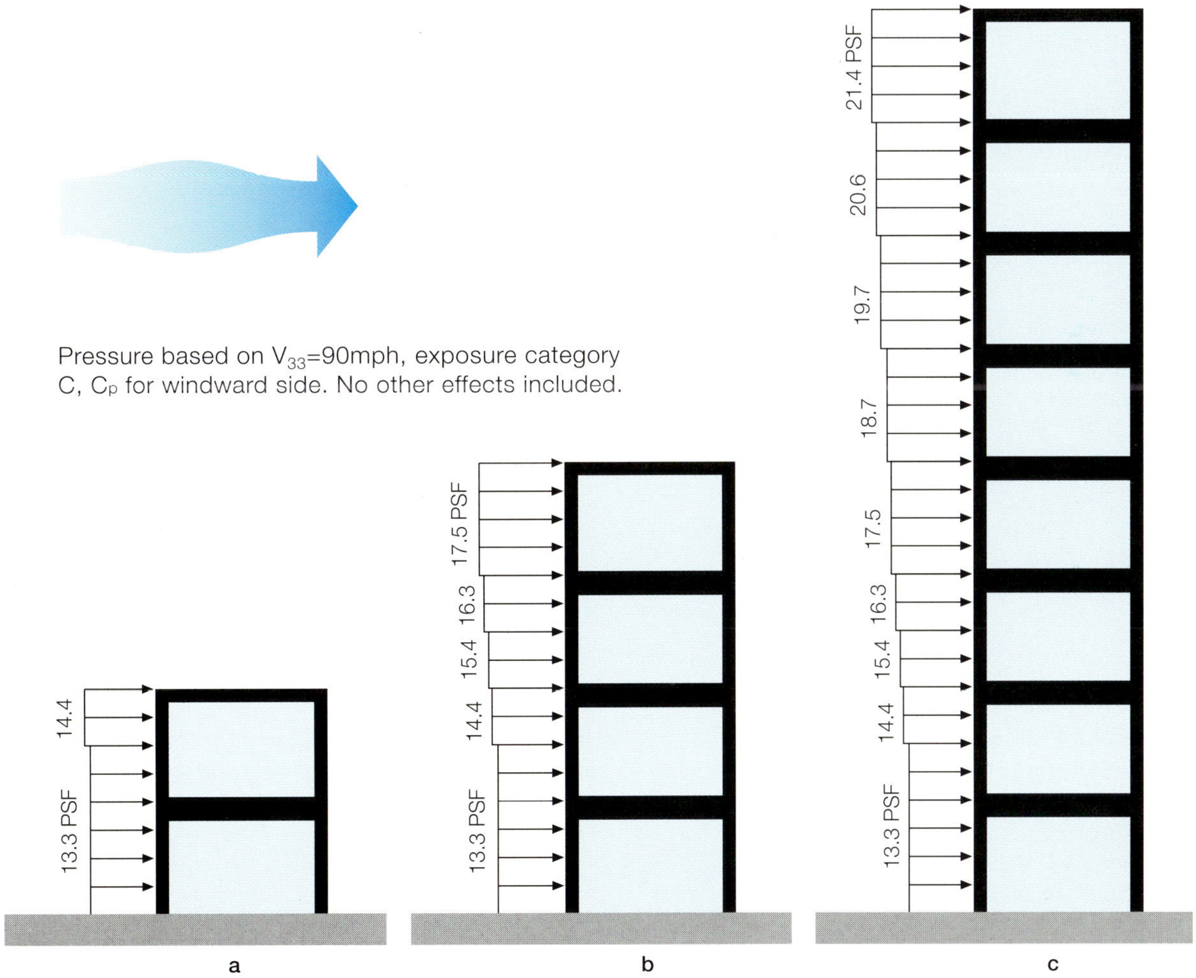

Some references even provide applicable pressures for the increments. Wind pressure then is assumed to be a constant value for each height increment, as is indicated in Figure 23. In these diagrams, the building story heights happen to coincide with the pressure step increments, either at the story level or at mid-story height.

Practical considerations arise when a stepped pressure profile is used for buildings of other story heights. Figure 24 illustrates stepped-pressure profiles for a building in which the floor-to-floor heights do not coincide with the steps of the profile. The first diagram shows the pressure profile in stepped increments that would result from using standard steps provided in the references. The pressures at each story vary, and thus present a practical problem–not insolvable, but inconvenient.

When wind forces are determined for a building's main wind-resisting structural system, procedures of statics must be used to obtain the correct force magnitudes. This is computationally straightforward when pressure steps match story levels, because the stepped profile distributes a uniformly-distributed pressure on each story. Treatment of this situation is the same as that of a single uniformly distributed load on a beam. When the stepped pressures occur randomly, force calculations are more complicated.

Figures 24b and 24c illustrate alternative pressure profiles that simplify the problem that arises in Figure 24a. Using calculated pressures at story or mid-story levels (another application of the power law), simplified pressure profiles can be obtained. A stepped profile similar to that of Figure 24c is especially convenient when lateral forces are determined at each floor level and the enclosure wall is framed vertically between stories. When preparing stepped-pressure profiles, stepped pressures should be taken conservatively; in other words, don't use the lower pressure value for a step increment.

Windward Wall Design Pressure

The windward wall pressure that is ultimately to be used for design purposes includes both an external pressure (provided above), plus either an internal pressure or consideration for the effect of an internal pressure. The pertinent equation has the following form and distinctly shows the external and internal portions. Equation 5 shows the final form for determination of the main-frame windward-wall design pressure, as provided in ASCE-7 references:

$p = q_z GC_p - q_h(GC_{pi})$ **Equation 5**

Observe first that the design pressure is denoted by the same symbol p that has been used throughout this monograph to indicate pressure. Therefore, the context of use of the term p must always be clear.

Note that for the internal pressure portion of Equation 5, the term GC_{pi} is enclosed in parentheses, whereas the terms stand alone in the external pressure portion. This is because a single coefficient is provided for GC_{pi} that includes consideration of building openness and consequent internal pressure effects, as well as gusting effects.

This is a simplification based upon empirical test data obtained using both building openness and gust effects as variables and is the sort of procedural variation that makes the equations appear to be more complicated than the concepts. Nonetheless, it is the way the references are organized.

Two values of the GC_{pi} coefficient typically are prescribed: one value with a positive sign and another with a negative sign. The vector directions of these pressures are in accordance with the convention that pressure acting toward a surface is positive.

Standard procedures make no distinction in the placement of openings. Instead, a building is simply to be categorized as fully open, partially open (as described earlier), or as "other" if neither of the first two classifications apply. A fully closed building is not included in these categories, although the third category approaches that condition.

Finally, with regard to Equation 5, the negative sign between the two terms on the right side is a correct algebraic representation of the pressure vector which correlates with algebraic signs attached to external and internal pressure coefficients. Consider the windward wall. The external pressure coefficient acts toward the wall and has a positive pressure coefficient. Two internal pressures are required to be considered–one positive and one negative. Looking only at pressure acting away from the wall (on its inside face), with a negative pressure coefficient, the net pressure on the wall is the sum of the two pressures–an exterior pressure and an interior pressure acting in the same direction.

Algebraically, Equation 5 equates to the same thing. The first term is positive; the second term is the negative of a negative value, which is positive, and so the pressures are added algebraically to produce a design pressure.

Leeward Wall and Side Wall External Pressures

Procedures for determining pressures for leeward and side walls are based upon data somewhat more empirical than the data for windward wall pressures; i.e., coefficients are based upon a simplified set of variables. The general equation for these pressures (it is the same for both leeward and side walls) has the same appearance as the previous equation for windward wall pressure. The subscripts change, however, and imply that parameters are different. The equation form is:

$p_h = q_h G C_p$

The subscript *h* indicates that a height different from *z* in the windward wall equation is used–specifically, *h* is the average (mean) roof height above the ground surface for the building.

Observe that external pressures for leeward and side walls obtained from the equation given above are a constant value for the full height of the building, whereas pressure varies with height for the windward surface. Hence, the pressures produce a uniformly distributed loading on the leeward wall and on each of the side walls. This is demonstrated for the leeward wall in Figure 25 showing buildings of different heights.

It is noteworthy that the leeward and side wall pressures for buildings of several stories have the same value at each story and do not vary from floor to floor. The validity of this approach has yet to be proven by researchers for some air flow conditions, but it is accepted procedure.

Leeward and Side Wall Design Pressures

The equation for determining design pressures for leeward and side walls has a form similar to that for the windward wall design pressure and has the following form.

$p = q_h G C_p - q_h (G C_{pi})$ **Equation 6**

To account for the possibility that internal pressure

Figures 24a, 24b, 24c VARIATIONS OF STEPPED PRESSURE PROFILES

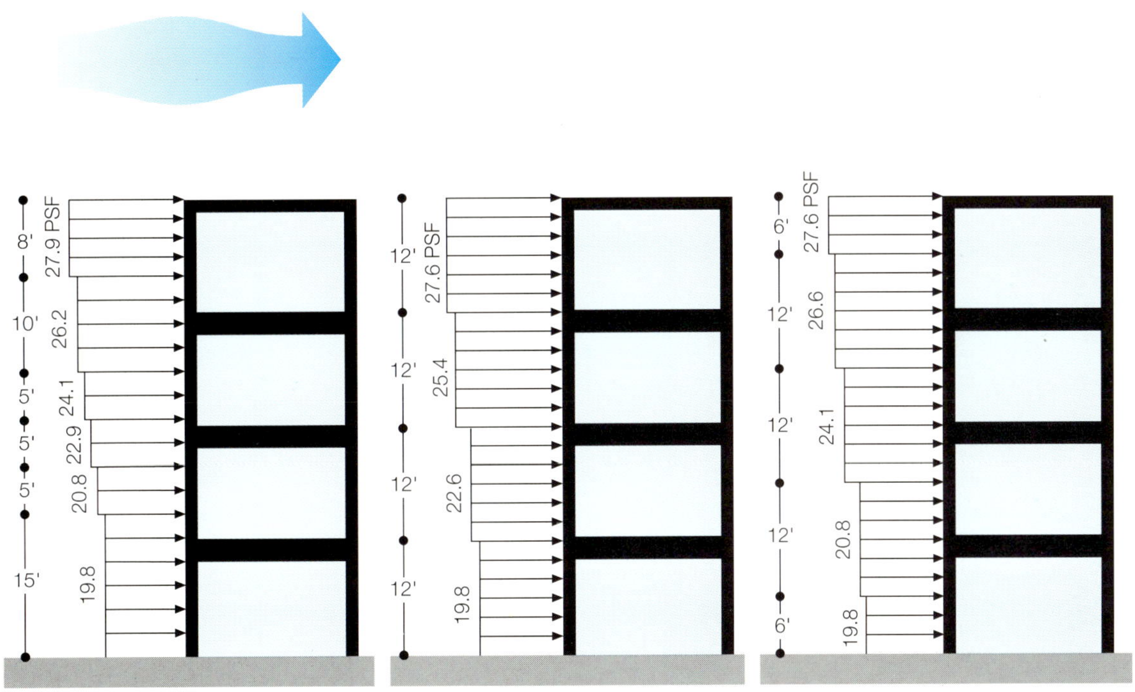

a
Stepped at standard height increments

b
Stepped at story heights

c
Stepped at mid-story heights

may be either positive or negative, current procedures for leeward and side walls require that pressures be determined for both conditions.

Roof Design Pressure

The equation for roof design pressure has the same form as Equation 6 for leeward and side walls, including the same subscripting. In other words, only the pressure coefficient for the roof changes. The equation, in working form, is as follows:

$p = q_b GC_p - q_b(GC_{pi})$ **Equation 7**

Main-Frame Design Pressures

Figure 26 shows a summary of overall wind design pressures and pressure equations that are considered for a building's main-frame. The pressure equations use symbols in accordance with ASCE 7-93 and ASCE 7-95. Other references use somewhat different forms and symbols.

There are two possible pressure magnitudes for the roof windward surface, depending on its slope. There are also two possible design pressure magnitudes for leeward and side walls–in accordance with the (GC_{pi}) term and four possible design pressure magnitudes for the roof when windward and leeward surfaces of a sloped roof and the (GC_{pi}) term are considered.

Our discussions and figures illustrating pressures pertinent to design of a building's main-frame have not considered ceilings and their effects. A ceiling under a roof is not believed to alter the development of internal pressure at the underside of a roof surface. The main-frame design pressures remain as described herein, whether or not a ceiling is present. However, when

Figures 25a, 25b, 25c LEEWARD WALL PRESSURE PROFILES

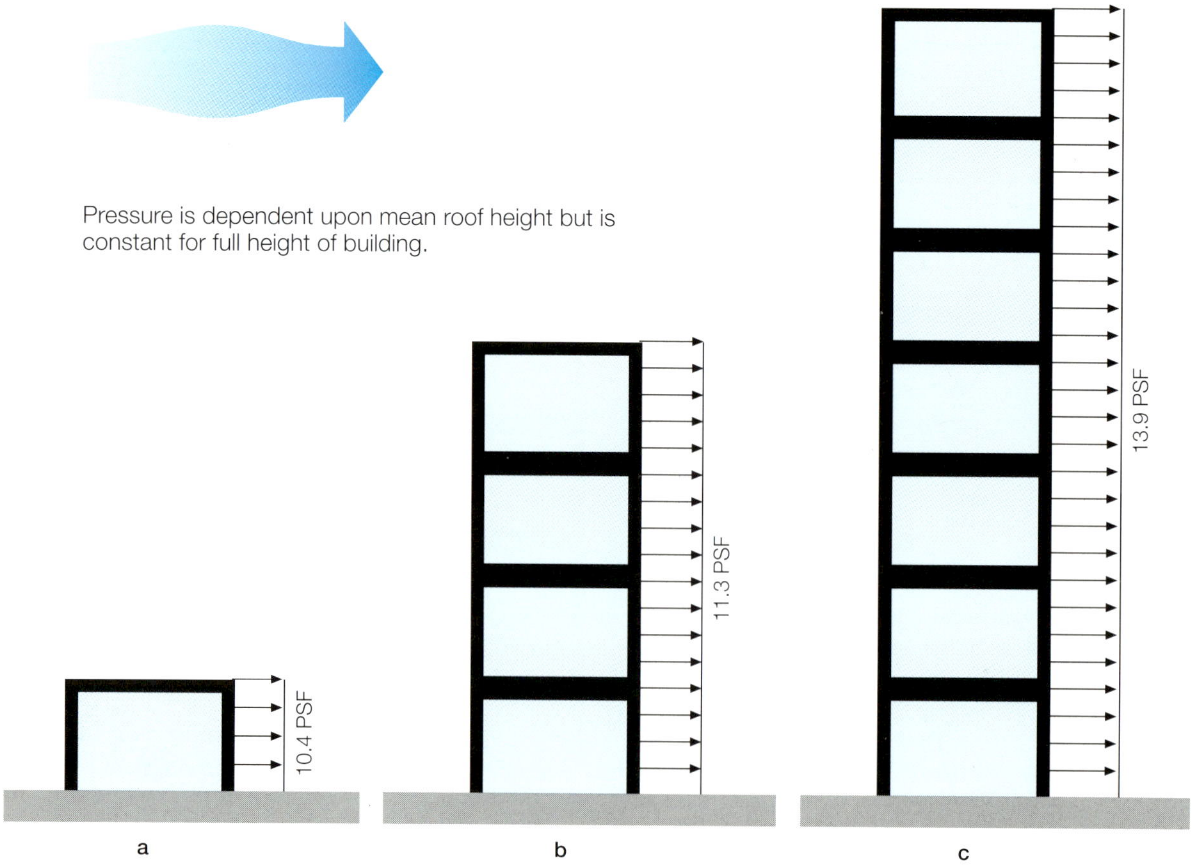

a pressure differential is possible on opposite sides of the ceiling assembly, internal pressure for the ceiling and consequent forces on other members should be considered in the design of that element.

Although pressures are shown acting on external surfaces in the schematic diagrams, it should be understood that some portion of the pressure is internally induced. Nonetheless, Figure 26 is helpful in providing an overview of all pressures that are taken into account for wind resistance by the main-frame.

One final comment regarding overall wind pressures: Notwithstanding calculated design pressure values, in no case is a pressure of less than 10 psf applied to a surface. This minimum design value is the same in all standard references.

Practical Procedures for Components and Cladding

The other principal phase of any wind analysis for a building considers wind pressures for the design of components and cladding. Recall that portions of some surfaces experience pressures that exceed those average pressures used for analysis of the main-frame. While such increased pressures will not substantially affect overall wind forces on the main-frame, they may be excessive for a particular component and location. This other analysis phase aims at ensuring proper performance of components in vulnerable locations.

The accepted procedures for determining component and cladding pressures is straightforward in concept. Design pressures for surfaces in certain locations are determined from the following general equation:

$p = q_h(GC_p)$ **Equation 8**

The term q_h is as described for Equation 6. Values for the (GC_p) term typically are provided in tables for particular component locations. Values are dependent not only upon the component and location, but also on the area of the surface, and therefore limiting dimensions apply.

Figure 27 provides a summary diagram of areas where localized pressure increases are to be considered. Note that the surface areas identified in this diagram occur at corners and roof edges. These are vulnerable surfaces which were discussed previously. The numbers identify areas which have different (GC_{pi}) coefficients based upon different conditions.

Unlike overall pressures which are considered for the main-frame and must be determined for the four possible wind directions, localized pressures on components are simply evaluated at each location. The worst-case wind effect has been taken into account in the GC_p coefficients.

BUILDING DESIGN CONCEPTS TO RESIST WIND

The preceding discussions have provided a detailed review of wind-building behavior and the procedures for determining magnitudes of pressure on building surfaces.

This section addresses the building structure and how it may be designed to resist these pressures. After

Figure 26a

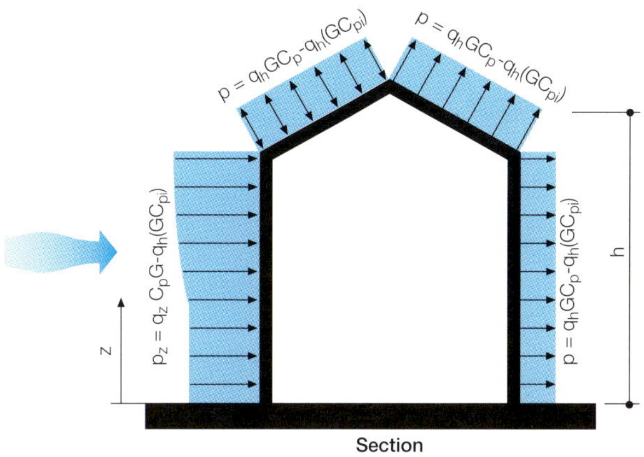

Figure 26b

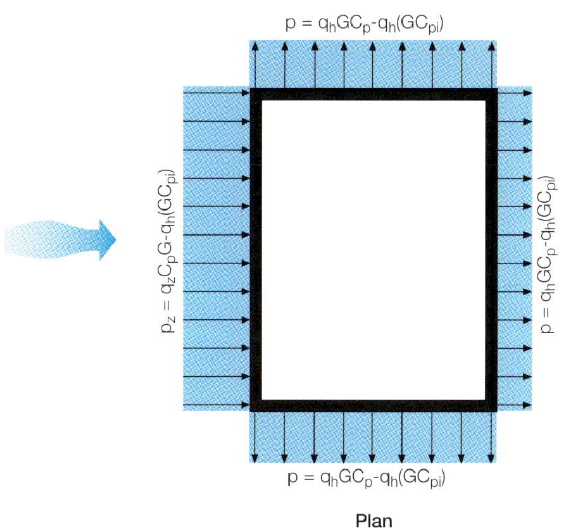

basic concepts are presented, structural systems of two buildings are examined in detail. Since concepts are emphasized here, the examples are examined qualitatively rather than numerically.

Assembled Parts

Any building can be considered as an assembly of parts or components. This concept is valid even for so-called "monolithic buildings" (explained later) and is therefore a primary consideration in designing or evaluating buildings to resist the effects of wind.

These parts may be major structural elements or seemingly non-structural elements that contribute to the enclosure of space. They all play an important, but

Figure 27 COMPONENTS AND CLADDING—AREAS OF LOCALIZED PRESSURE INCREASE

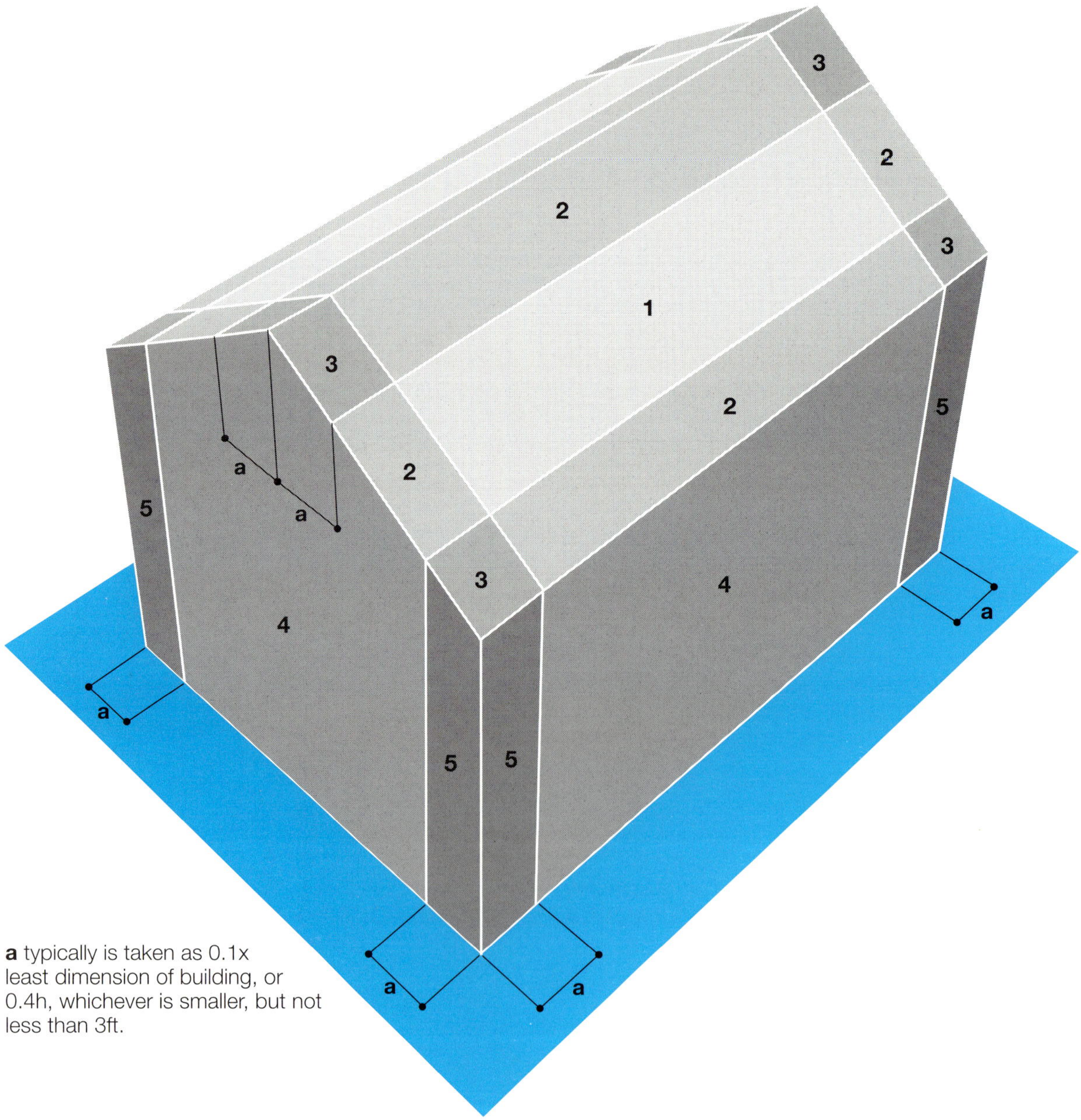

a typically is taken as 0.1x least dimension of building, or 0.4h, whichever is smaller, but not less than 3ft.

different, role in the manner in which the building resists wind.

If a building is essentially monolithic, such as one of poured-in-place reinforced concrete, certain sections still perform specific tasks, and it should still be considered as an assembly of parts.

For example, a roof deck is distinctly different from a foundation wall.

Components vs. Main Wind-Frame

Once the assembled parts premise is understood, it then becomes necessary to distinguish which parts (or group of parts) become the primary wind-load resisting system, and which parts are non-structural and serve only to enclose space.

The primary or "main" wind-force resisting system is an assembly of structural elements assigned to provide support and stability for the overall structure. The main system generally receives wind loading from more than one surface (multi-directional). All other elements that are not a part of the main wind-force resisting system comprise a different design category, frequently referred to as components and cladding. It is useful to picture the components (windows, wall panels, cladding and the like) as barriers in the path of the wind. These components develop forces that must be transmitted to the primary structural system.

The importance of connections (between components and to the main frame) cannot be over emphasized. The majority of wind damage in buildings can be attributed directly to the failures in these connections. This is particularly so in a hurricane wind. Hurricanes produce persistent wind forces, acting in multiple directions, that work on components, trying to pry them loose. Once a component is loosened, the wind gets a hold, so to speak, and progressive failure usually follows.

Figure 28a WALL COMPONENT EXAMPLE

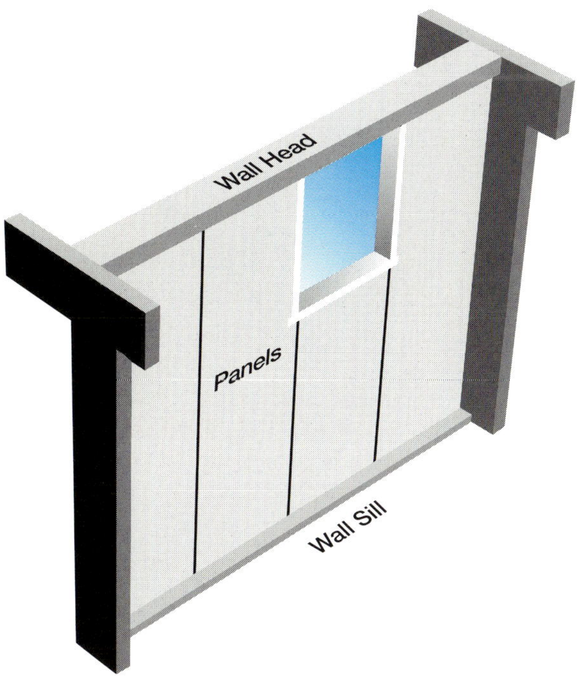

Figures 28b, 28c WALL COMPONENT EXAMPLES

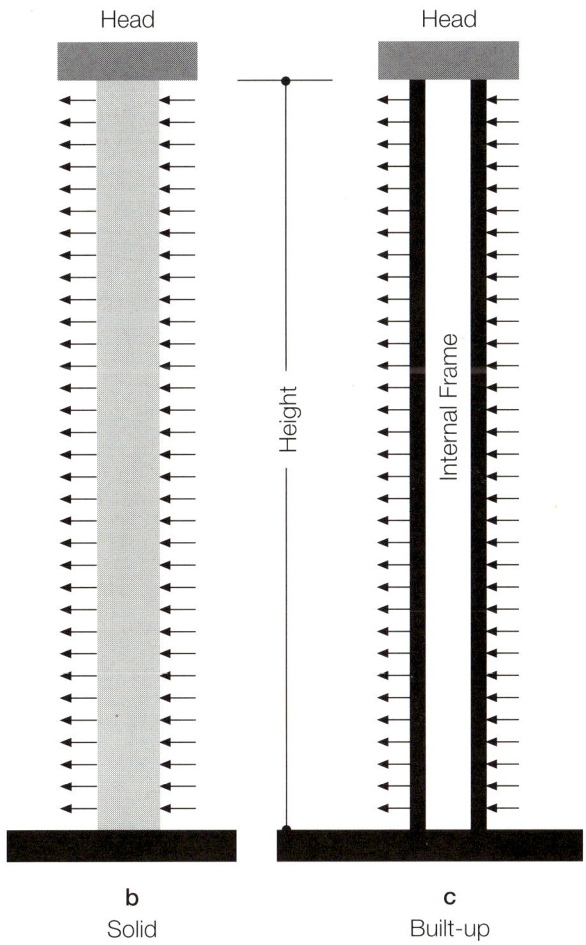

Figure 29 POST-AND-BEAM (VERTICAL FORCES ONLY)

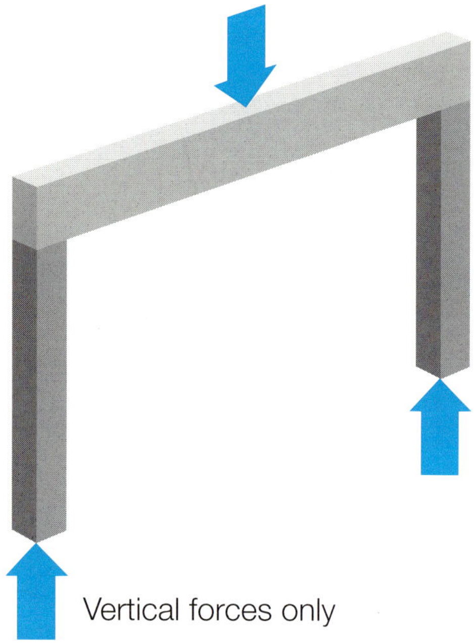

Example of Components and Loads

The magnitude of forces in component connections will vary, depending upon their composition and their placement within the building system. Assume a wall panel, as shown in Figure 28a is placed in the path of a wind that is perpendicular to the wall. Assume that the panel is attached to another component at the sill and head. Two cases are described: a solid homogeneous panel, shown in Figure 28b, and one that is "built up" as shown in Figure 28c.

The solid panel must have enough strength to span vertically the distance from the sill to the head, carrying pressures on both sides of the panel. These pressures might be additive, perhaps if the inside surface has to resist pressure inside the building, and the outer surface experiences a suction on the leeward side of the building.

The intensity of these pressures would depend on the factors discussed previously, including component pressure coefficients. The connection at head and sill must be strong enough to develop and transfer the total pressure force (pressure times panel surface area) to the head and to the sill. In this case it would be one-half the total force to each distributed over its length.

Figure 30a RIGID FRAME (VERTICAL AND HORIZONTAL FORCES)

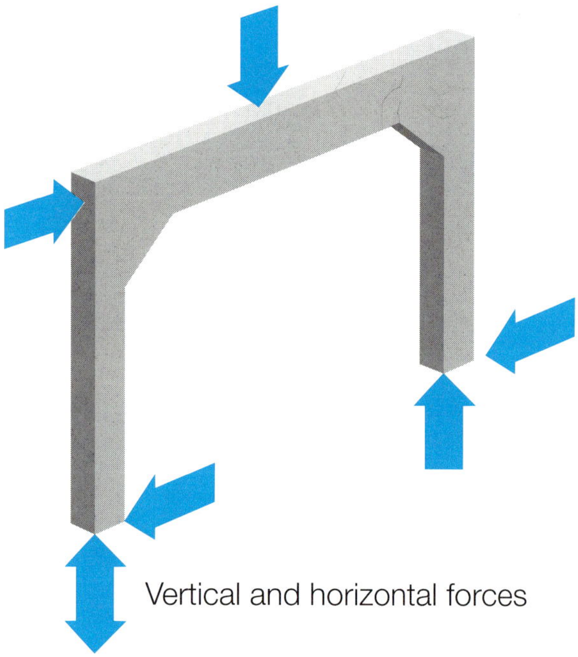

Figure 30b BRACED FRAME (VERTICAL AND HORIZONTAL FORCES)

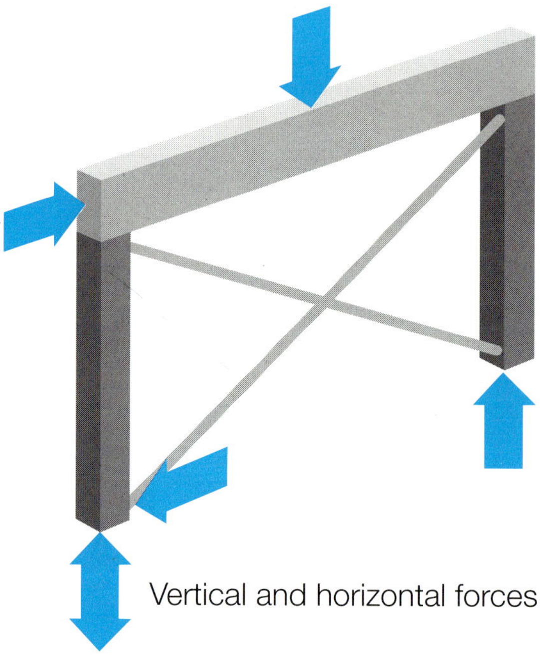

Compare this force action with the force action on the built-up panel shown in Figure 28c. The cladding on one side needs only to transfer the pressure on its side to the internal frame. The same can be said for the other surface, however, the head and sill connection must handle the accumulation of both sides, in the same manner as the solid panel.

The wall in Figure 28a also shows a window. This window would perform much like the solid wall panel with additive surface pressures summed up over its area and transferred to the window frame. The window glass must be strong enough to withstand the pressures, and the window frame attachment to the supporting structure must also be adequate. Otherwise the glass, at the very least, would be destroyed, leaving an opening in the wall, allowing wind to enter the building.

The head may be an eaves beam, spanning horizontally between two main frames and may receive forces from several panels and windows. Some refer to this element as a collector. This eaves beam's connection to the main frame must be adequate for the accumulative action of all the components attached to it. Forces in this connection then would become one of the loads on the main frame.

Primary Main Wind-Frame Systems

A primary wind-load resisting system (main-frame) is any structural configuration capable of resisting horizontal forces.

Some structural systems, but not all, used to support gravity loads are also capable of resisting horizontal loads.

Vertical Elements of Main-Frame Systems

Vertical elements in a lateral-force resisting system must have the capability to transmit horizontal forces downward, ultimately to the foundation system. A post-and-beam system, as shown in Figure 29, cannot resist horizontal forces. However, the rigid-frame system, shown in Figure 30a, can resist both vertical and horizontal forces and can be used as a vertical element.

A post-and-beam system can be modified to resist lateral forces by the addition of bracing, as shown in Figure 30b. The resulting braced-frame system can function as a vertical element in a lateral-force resisting

Figures 31a, 31b, 31c BEARING AND SHEAR WALLS

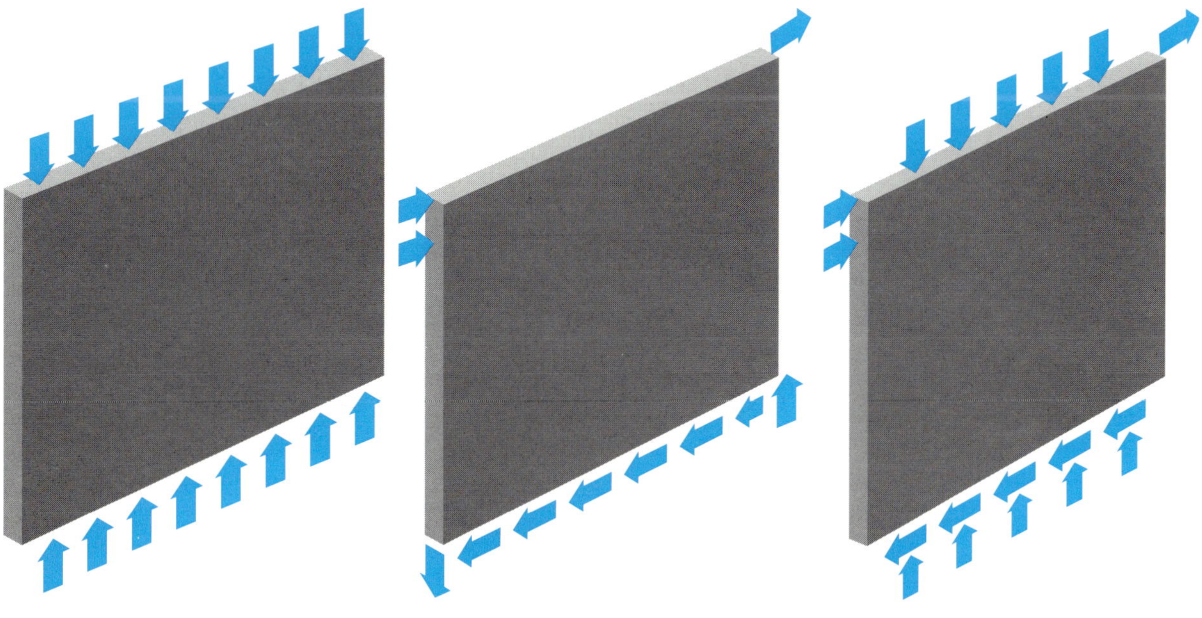

a BEARING **b** SHEAR ONLY **c** BEARING AND SHEAR

Figures 32a, 32b, 32c HORIZONTAL ELEMENT EXAMPLES

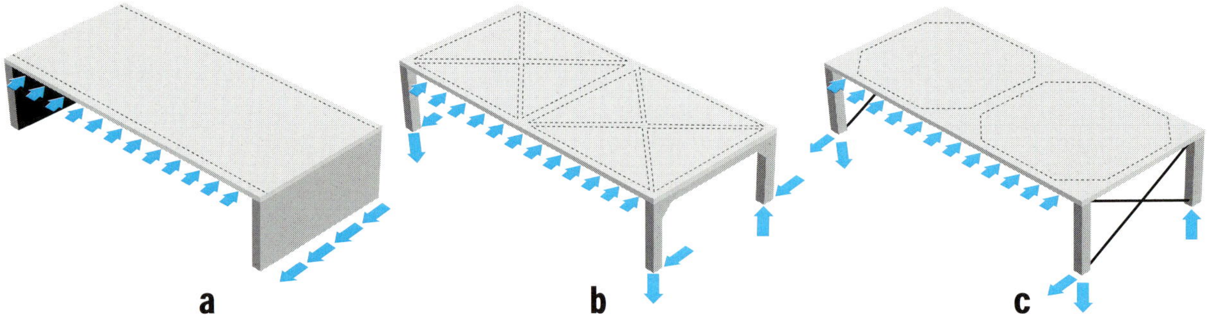

Figure 33 STABLE MAIN-FRAME STRUCTURE

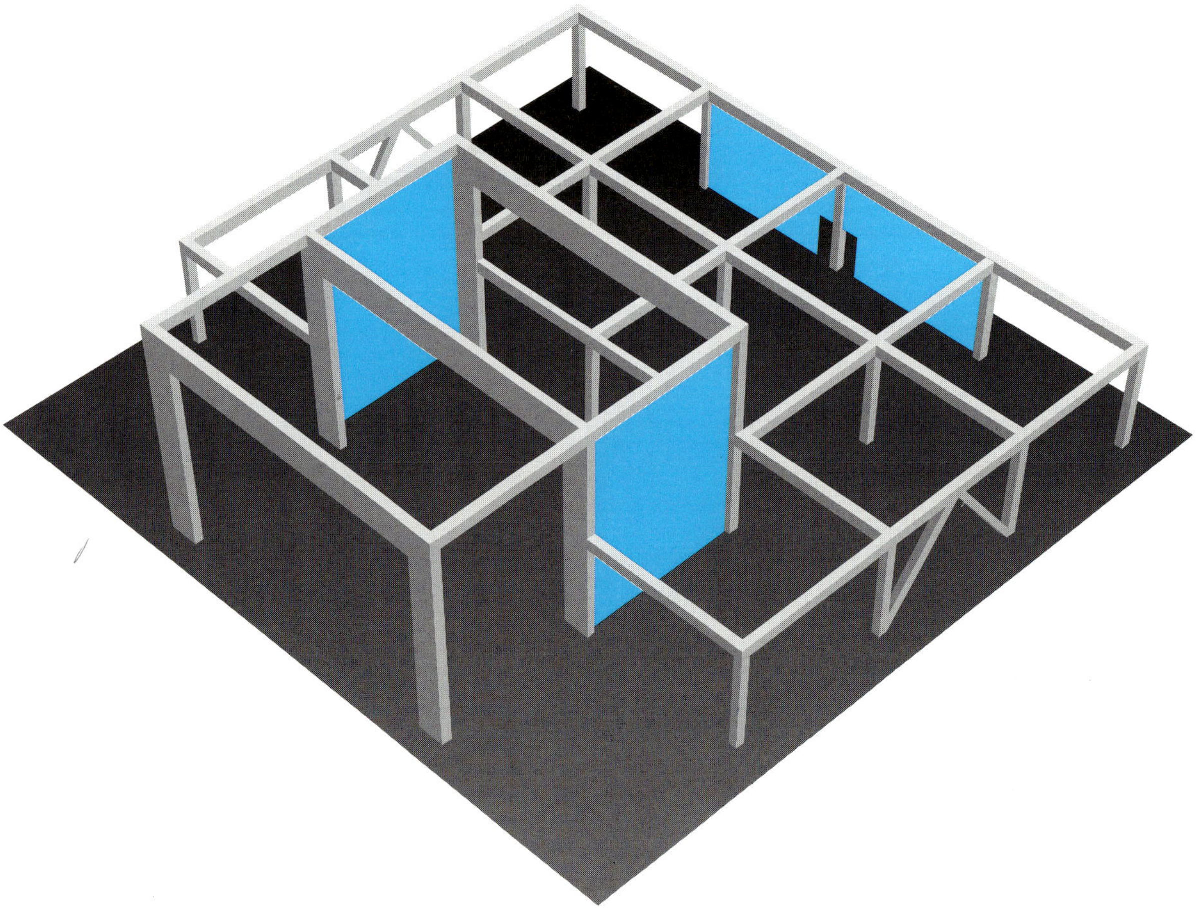

system. Observe, that with the type of cross-bracing shown, the bracing elements come into play only when a horizontal force is applied.

The third type of vertical element is the shear wall as shown in Figure 31. A shear wall is a planar element as contrasted with linear elements. Three walls are shown, one bearing, one shear, and one combination bearing and shear. Note that a shear wall need not be a bearing wall. A non-bearing shear wall can be a very useful element and is frequently used in buildings to develop wind shear forces only. However, the combination wall functions much better because of the stabilizing effect of the bearing forces.

Horizontal Elements of Main-Frame Systems

The main-frame system must consist of both horizontal and vertical elements. Horizontal elements need not be in a horizontal position as long as they can transfer lateral loads to vertical elements. Horizontal elements are almost always a diaphragm. A diaphragm is a surface structure functioning principally in shear, but which has chord forces developed around its perimeter.

Diaphragm construction may vary from a monolithic form to an assembly of pieces. Figure 32 illustrates three different systems.

The diaphragm in Figure 32a is subjected to a load along its eaves and transfers them to the two shear walls at its ends. The diaphragms shown in Figures 32b and 32c are not true diaphragms as defined above, but they are classified as such because of their similar purpose. The diaphragm in Figure 32b transfers its eaves load to rigid frames at its end, while the one in Figure 32c transfers its eaves load to braced frames at its ends.

Figure 32b works more like horizontal cross-bracing, while 32c has essentially horizontal rigid frame construction. Even though they are constructed differently, the codes accept them all as diaphragms.

The examples shown in Figure 32 are incomplete in that they are unstable in the direction perpendicular to the eaves forces. They are shown in this manner simply to illustrate a complete main-frame in one direction only.

Summary

The main-frame system must be three-dimensional. This usually means that it must be complete in two directions. Each direction must consist of both horizontal and vertical elements. These elements may be rigid frames, braced frames, shear walls and diaphragms. They can be used in almost any combination, but should be located in a building in such a pattern that gives complete stability with as little torsion as possible. One way to achieve this is to use the elements symmetrically, wherever possible. The term "symmetrically" here refers to stiffness, and not necessarily type and shape of the system. It should also be noted that vertical elements can be located in interior walls as well as the exterior walls.

The implementation of these basic structural principles into any building design rests entirely with the understanding and creativity of the architect. Figure 33, for example, represents a possible building structure. Both the high and low roofs must be designed and constructed to function as diaphragms in order to carry the wind loads to the vertical elements. The vertical elements for the high structure are the rigid frame and shear walls located as shown. The low structure has one shear wall (with an opening) in an exterior wall, two braced frames, and it also makes use of the lower part of the shear walls of the high structure. In fact, if the shear wall of the high structure, along with its connections and details, were strong enough, bracing of the two frames could be eliminated. Clearly, other wind-resistant structural solutions for this configuration are also possible.

DESCRIPTIVE EXAMPLES OF BUILDING DESIGN FOR WIND

Example Building Types

There are many possible construction types and details to be considered when the design process focuses on the effects of wind. This section describes some of these alternatives in relation to specific building examples. Two different building types are shown and discussed. One example is a three-story apartment-type building with a flat roof. The other is a large, single-span steel-framed building with sloped roof surfaces such as might be used for a variety of industrial or commercial purposes.

Schematic diagrams of these two building types are shown in Figures 34 and 35. These examples are also used later in this monograph, at which time more specific construction details and calculations are provided.

Procedures Common For Both Buildings

Determination of Velocity Pressure

The velocity pressure should not be confused with

design pressure. Velocity pressure, sometimes referred to as stagnation pressure, is the equivalent static pressure (in psf) from wind, considering all physical effects except any influenced by the building.

Neither building example is a "flexible-type" nor are they "open-type" (see ASCE 7 references for classification), so these two considerations may be eliminated. Both buildings, having heights of less than 60 feet, qualify for analysis by a special procedure specified in ASCE 7-95 (which is discussed in the numerical calculations section). Discussion for each example depends on the magnitude and direction of wind pressure on building surfaces. Recall that these are collected by components of framing elements which then transfer the loads to the main-frame.

In order to obtain the velocity pressure, begin by establishing the basic wind speed *V* in miles per hour. This basic wind speed is determined either by the geographic location of the building and the specific building code being used, or it is extablished by the building official.

The basic wind speed is next modified according to its exposure category (A, B, C or D) and for specific heights of the building surfaces. From this data, the stagnation pressure can be determined either by formula or reference tables. The formulas use a multiplication factor K_z to account for heights and exposure categories. The stagnation pressure is then modified by multiplying it by the importance factor *I*, based on the building's use. The pressure might be modified once more by a gust factor *G* and/or a topographic factor K_{zt}. The final velocity pressure may have several values, each relating to the various surface heights.

Distinguishing Main-Frame and Components

Procedures to this point are the same for both building

Figure 34 APARTMENT BUILDING EXAMPLE

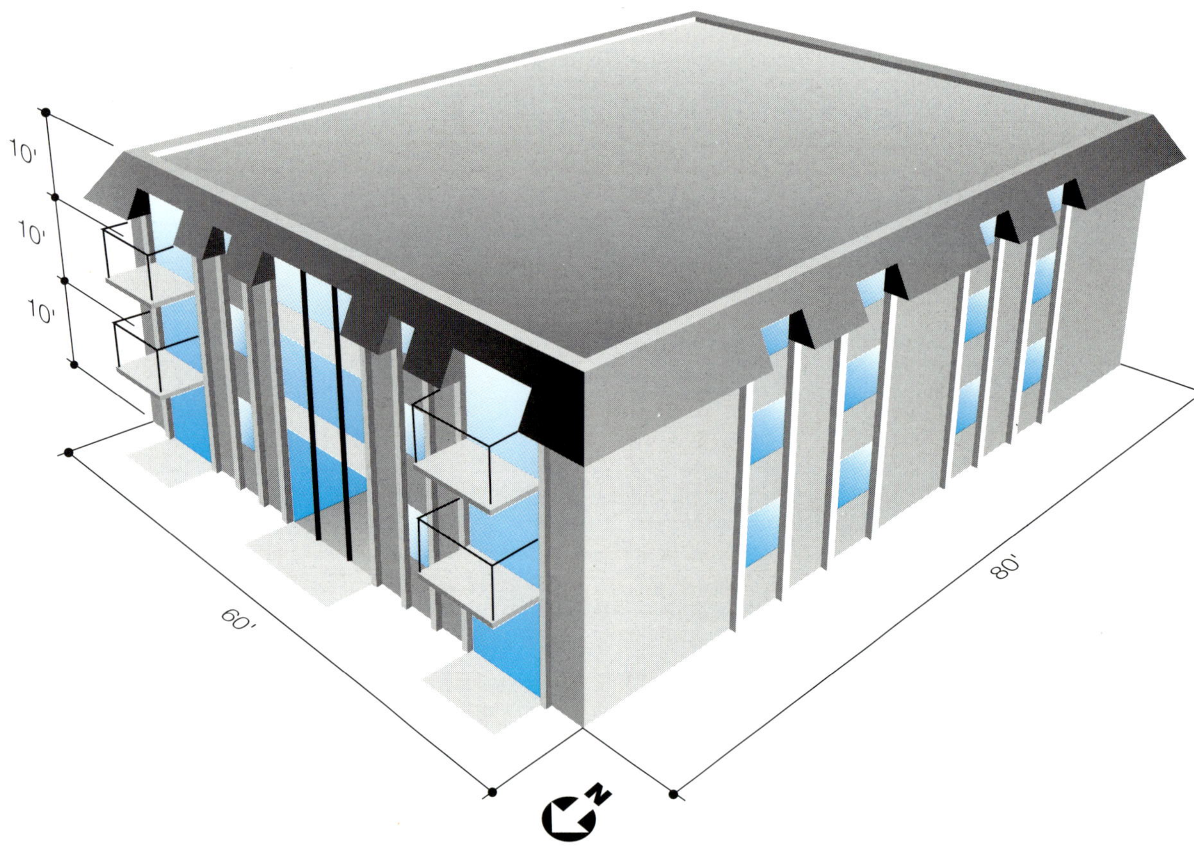

examples and for all main-frame and components parts. Some components may also serve the function of collectors, transferring loads to the main-frame. Procedures in the analysis depend next on whether they pertain to the main-frame or to the components.

For example, all wall sections shown in Figures 36, 37, 38 and 39 show an inward pressure and an outward pressure (suction). The value of each pressure depends upon whether the wall is being considered as a component or as a collector for the main frame. The value of a pressure (acting inward or outward) for design of components and cladding may be 1.5 to 2 times the value of the pressure used in the design of the main-frame. Recall that those building surfaces near edges and corners will experience higher wind pressures. These higher pressures act over a relatively small area of a surface, but must be considered in the design of the component and its connections. This is not the same as the load that is transferred to the main-frame. Once the component and its connections are designed in the areas of increased pressures, they are normally not adjusted downward for other surface areas. However, refinements in this process vary widely among designers.

Wall pressures that are shown and referred to in the example building discussions are design pressures calculated by using pressure coefficients, either for components or for the main-frame, as is appropriate.

We emphasize that a collector, as used in this discussion, refers to a building element that collects and feeds loads to the main-frame system. This distinction provides a more complete picture in development of the load path as pressures are aggregated to produce forces. When a collector also functions as a component or cladding, it must be designed for any pertinent increased pressures, even though it is not transferring them to the main-frame.

Observe that each consideration is incorporated

Figure 35 SINGLE-SPAN STEEL FRAME BUILDING EXAMPLE

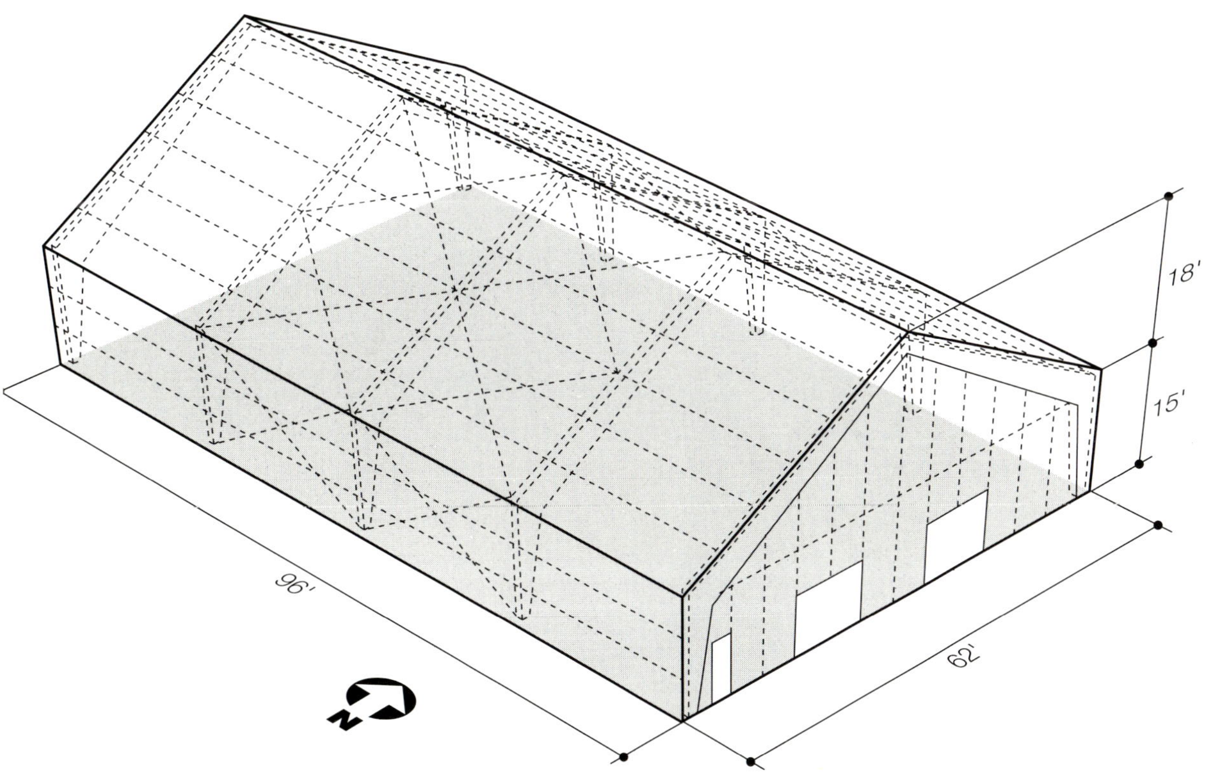

through a multiplication process. Finally, the end result is a design pressure value in pounds per square foot (psf) and a direction in which the pressure acts on various building surfaces. This pressure must be at least 10 psf, but could be many times higher. This value also may vary over the surface or element under investigation. Either the maximum value is used over the entire surface, or an average value may be used if the surface area of larger pressure is relatively small.

Example—Apartment-type building
The first example is the three-story apartment-type building shown in Figures 34 and 58.

General Building Construction
The framing of the floors and roof consists of joists spanning in the east-west direction. These joists might be either steel or wood, and they could either span the full distance between the east and west exterior walls, or they could bear on interior partitions to reduce the span. The use of interior bearing walls would reduce the joist size, but larger joists spanning the full distance to the exterior walls would permit greater flexibility in the interior partition layout. Nevertheless, whichever is used, the east and west exterior walls are bearing walls, and require appropriate lintels over windows and openings. The north and south exterior walls and many, if not all, of the partitions would be non-load bearing.

The floors and roof require a structural deck spanning the distance between the joists. This deck could be constructed from a very broad range of materials–from poured-in-place gypsum or concrete to planks and panels attached to the joists and to one another.

The exterior masonry walls could be a veneer attached to a structural backing or they might be "solid" masonry. Each of these construction possibilities will be discussed with respect to the wind consideration. However, we will assume at the outset that the main-frame structural system, for this example, in both the long and short building directions, consists of horizontal diaphragms and vertical shear walls to resist wind forces. Observe that this is an overall description of the major structural elements which contribute to main-frame resistance.

Roof Components
We will first discuss components, beginning with the roof. The magnitude of the wind pressure has been calculated, and since the roof is flat, the direction will be upward. Whatever the roofing material is, it must be attached to the structural deck to resist the uplift. For example, single membrane roofing covered with a

Figure 36 SOLID MASONRY BEARING WALL

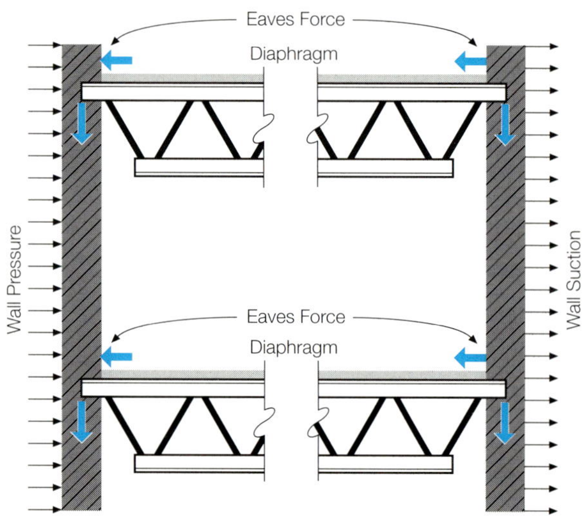

Figure 37 VENEERED BEARING WALL

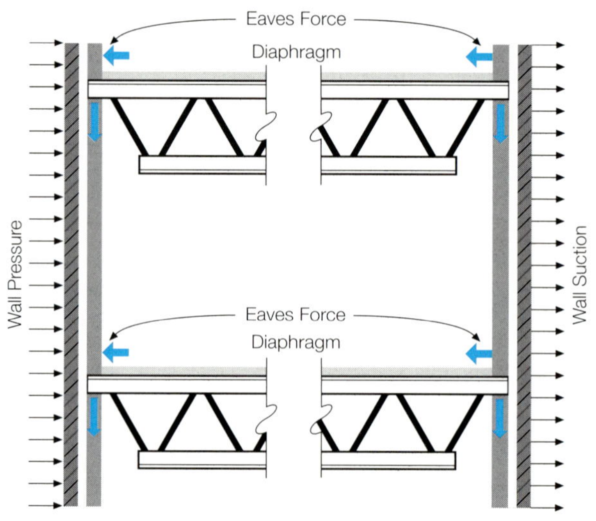

stone ballast will normally have a resistance of about 10 psf. A larger uplift pressure would require more ballast and/or mechanical attachment to the deck. The decking material must be attached to the joists to resist the uplift pressure in excess of its own weight plus the roofing weight. The roof joists must be anchored to the bearing walls to resist the uplift pressure on the surface it is carrying in excess of the dead load it is carrying. Note that live loads are not included to resist uplift forces.

Continuing with component considerations, observe that the floors are not affected by wind pressure (with regard to development of wind forces on components) and only participate with the main-frame system in resisting overall lateral forces from wind.

Wall Components–Windows

All of the exterior walls are components as well as collectors when considering inward or outward pressure on their surfaces. The window units must be selected on the basis of their capability of resisting the design pressure. If windows are located in areas of increased pressure, they must be able to withstand the higher pressure. The strength of the glass must withstand the pressure acting on its full surface, and the window frame must withstand the full window force (pressure times glass surface area). The frame must also be attached to the structural wall to withstand this same force. Responsible window manufacturers will provide this structural information in their catalogs.

Bearing Walls as Components–Solid Masonry

Consider the east and west bearing walls if constructed of solid masonry (Figure 36). The dead load carried by each wall (from the joists), plus its own weight, gives it a strong "bending strength" to carry the wind pressure vertically from floor to floor for either an inward or outward pressure.

If the wall tensile stress from this bending action exceeds the vertical dead load pressure, then vertical reinforcing will be required. There is also a pushing or pulling force between the diaphragm and the wall, often referred to as "eaves force."

Bearing Walls as Components–Veneered

An alternate to solid masonry walls is masonry veneer on a structural backing, as illustrated in Figure 37. The veneer does not participate in carrying the vertical loads from the joists. The masonry veneer must be attached to a backing to withstand the wind pressure normal to its surface. Usually, outward suction is more critical, and masonry wall ties are adequate.

The structural backing must span the vertical distance from floor to floor and withstand the wind

Figure 38 SOLID MASONRY NON-BEARING WALL

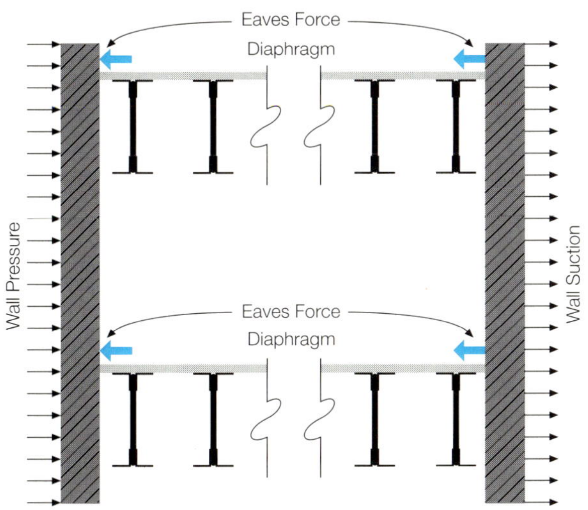

Figure 39 VENEERED NON-BEARING WALL

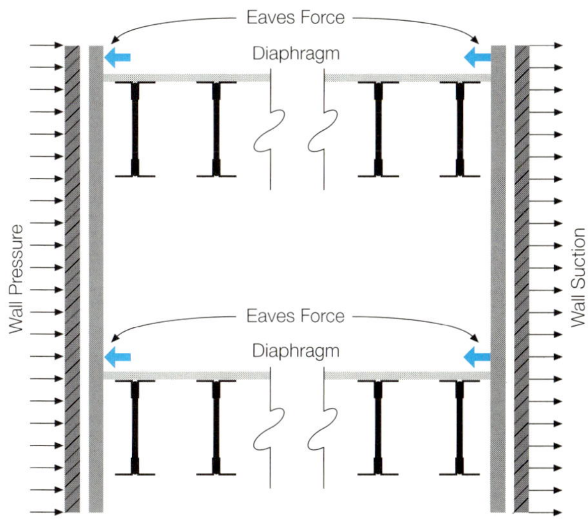

pressure normal to its surface.

This leads us to a new detail to consider. In both cases, the solid masonry and veneered wall must be attached to the floor or roof to withstand the accumulated wind pressure on its surface. Usually, outward forces are critical. This is the "eaves attachment" at each floor and roof level and is noted as an eaves force in both figures. Most codes require design for a minimum force of 200 lbs. per lineal ft. (plf) at these locations. This force is applied inward and outward and also parallel to the eaves. Note that in this example, the floors and roof are diaphragms supported by bearing walls, and there are no eaves beam collectors.

Non-Load-Bearing Walls as Components–Solid Masonry
Next consider the north and south non-bearing walls (Figure 38) with an inward or outward pressure perpendicular to the wall surface. A solid "non-bearing" masonry wall is not nearly as wind resistant as a load-bearing one. This is because it lacks pre-stress compression (pre-wind loading) resulting from the superimposed vertical load.

When examined for bending (caused by wind), a non-bearing wall has only its own dead load pressure to withstand the tensile bending stress when spanning vertically from floor to floor. When tension does occur (exceeding dead load pressure), reinforcing is required.

Non-Load-Bearing Walls as Components–Veneered
In a non-load-bearing veneered wall (Figure 39), the veneer's attachment to the structural backing is the same as in the bearing wall. The backing (without being bearing) must be capable of transferring its accumulated load to the eaves which must serve as a collector. This is accomplished by bending action and usually is not a problem. Also, the walls' attachment to the floor or roof eaves is the same as for the bearing wall but additional anchoring may be required at the sill (foundation).

Notice that the interior partitions do not function as components, although they may be load-bearing and/or be designed to function as shear walls and be part of the main-frame system.

In conclusion, it should be emphasized that the wind surface pressures for components frequently are much larger than on the same surfaces when designing the main-frame. Remember the effects of localized increased pressures. The numerical example beginning on page 68 illustrates this point.

Main Wind-Frame
The main wind-frame is not really a structural frame in a traditional sense in the apartment building example, but it is a lateral-force-resisting system serving the same purpose (Figure 40). The diaphragms and shear

Figure 40a DIAPHRAGM SHEAR WALL MAIN FRAME

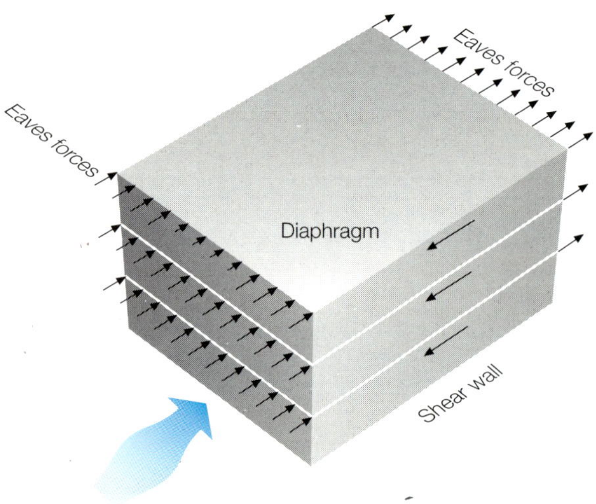

Wind toward non-bearing wall of Apartment Building Example.

Figure 40b DIAPHRAGM SHEAR WALL MAIN FRAME

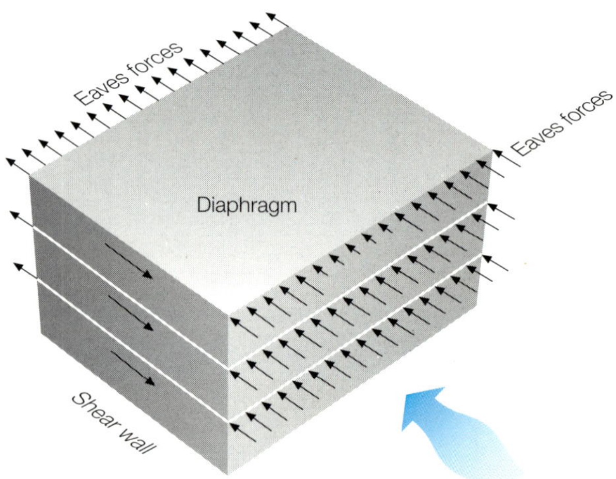

Wind toward bearing wall of Apartment Building Example.

walls are "surface structures" and must be properly constructed and attached to other elements to function adequately.

Architects, by their education and training, should be familiar with construction requirements for diaphragms and shear walls, and how they interact in a structural system. Diaphragms receive their wind loads from walls pressing or pulling against their eaves (edges). This load could be 200 plf as mentioned previously but could be greater as the wind wall loads are combined in calculations. The pressure on the windward side pushes the eaves in the same direction as the eaves pulling on the leeward side. In the simplified example of Figure 40, the eaves loads are indicated as uniform but not necessarily equal, and there is symmetrical loading. However, in many cases, especially when using procedures from newer codes, the resulting eaves loads take on a more irregular pattern, which induces torsion in the system.

Main-Frame Diaphragms

In this example, diaphragms collect and transfer horizontal loads to the shear walls. We first consider only the exterior walls (bearing and non-bearing) that serve

Figure 41 DIAPHRAGM OF LONG, SLENDER PROPORTIONS

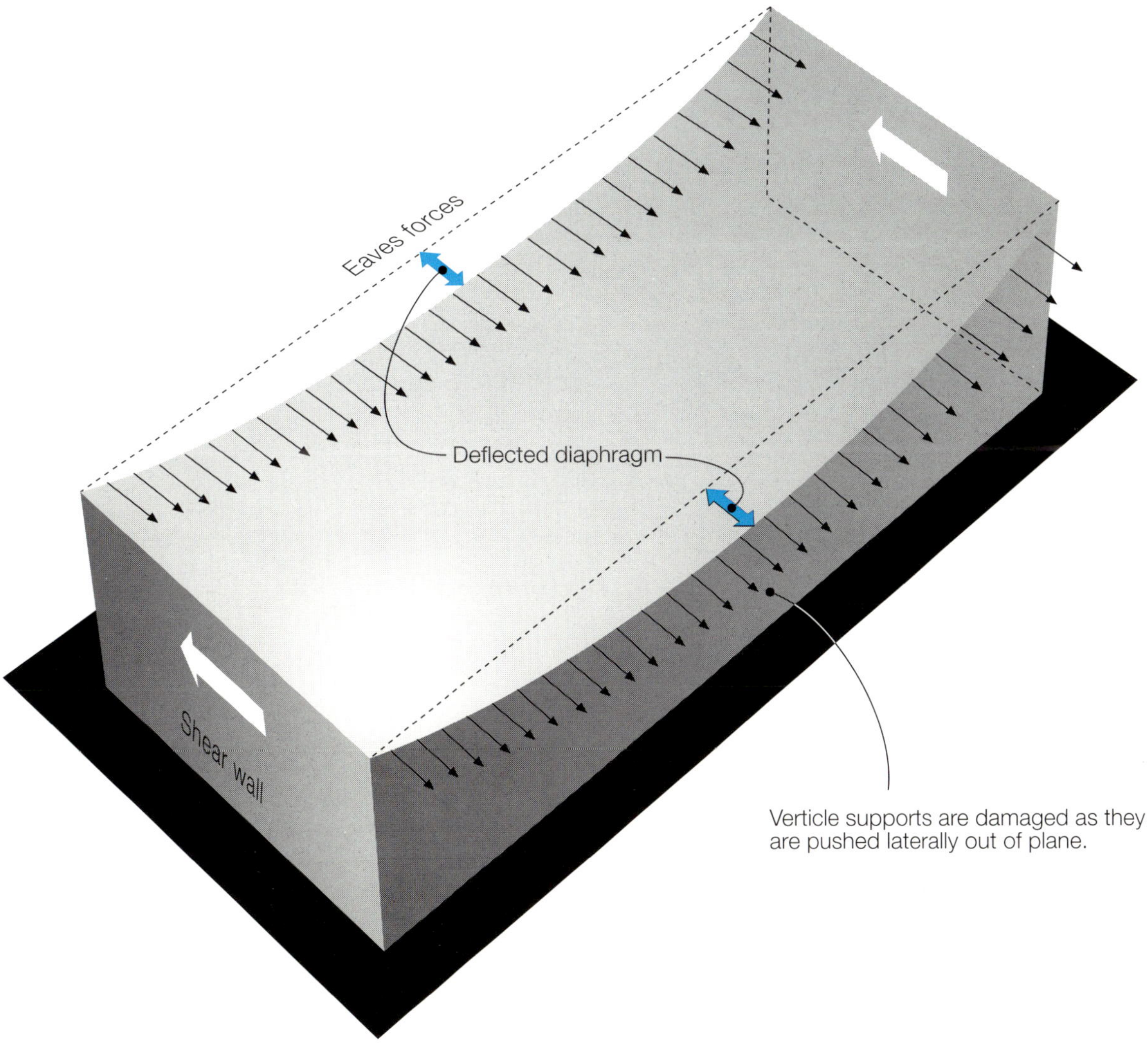

as shear walls. No interior shear walls are shown in either building direction at this point for the sake of discussion. (Later, in the numerical example, interior shear walls will be introduced.) Roof and floor decks constructed of monolithic material, such as gypsum and concrete, have an inherent capability to function as rigid diaphragms. The shear stresses are quite small (with proper diaphragm thicknesses), so each diaphragm needs only the chords around its perimeter. These chords could be in the slabs themselves, in wall bond beams, or provided by regular joists and rim joists.

It is much more difficult to construct diaphragms from planks and panels. Individual panels must be attached to one another so that they won't slide, and the attachments usually limit the diaphragm's shear capability as well as make the diaphragm flexible. These panel decks also require perimeter chords. Many safe-load tables giving shear values and deformation constants are available to help guide the architect in proper analysis and assembly.

In the example shown, the diaphragms should function well in transferring the eaves load to the

Figure 42 PROVISION OF AN INTERIOR SHEAR WALL

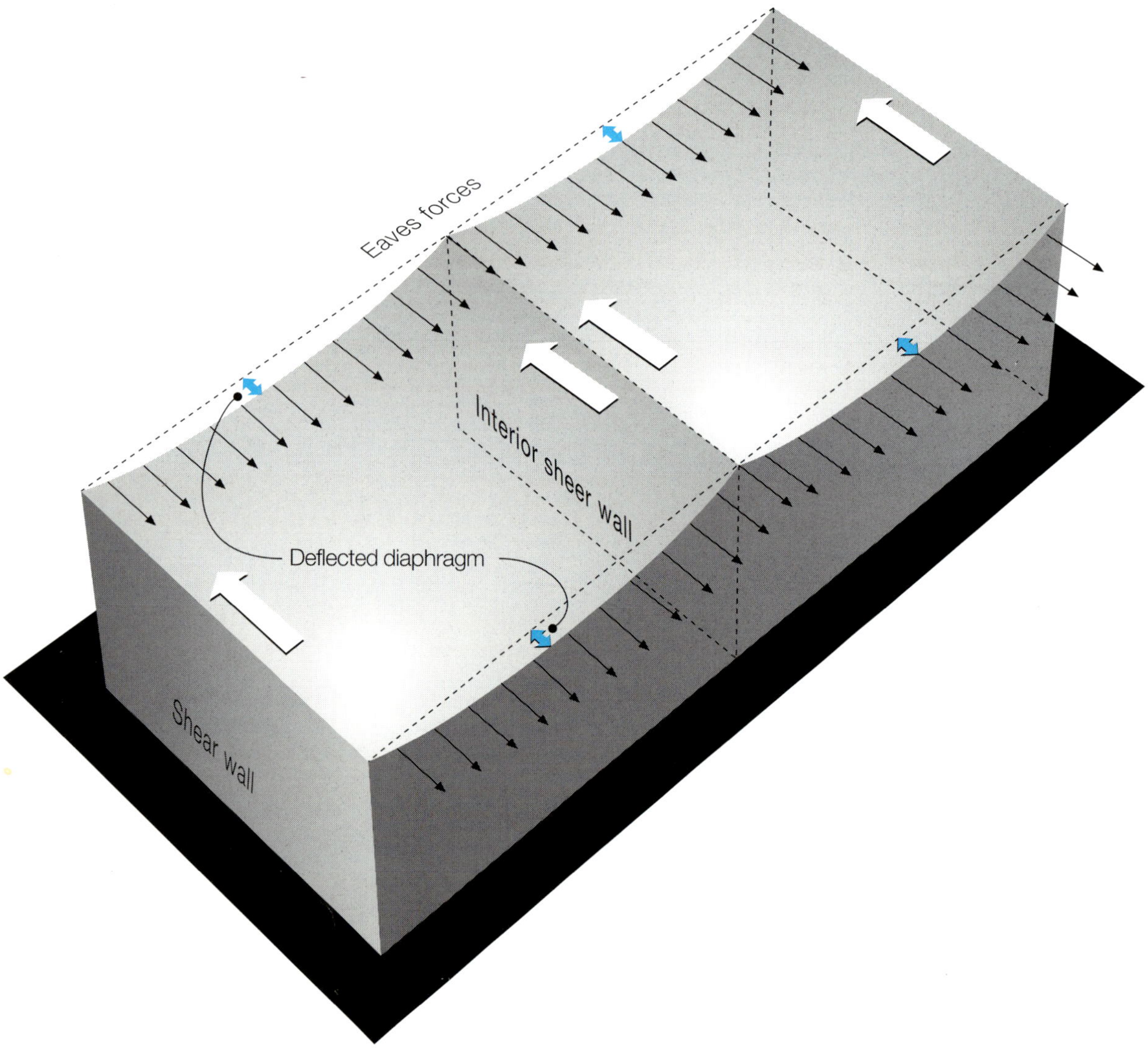

shear walls in either direction. This is because the proportions of the building in plan is only slightly rectangular. Diaphragm problems occur most often in buildings with long slender proportions (Figure 41).

The length-to-width proportion of a diaphragm is limited by most codes to 4 to 1. Even within these limits, a flexible diaphragm could deflect horizontally to such an extent as to damage brittle bearing supports. Figure 41 shows how the eaves load deforms the flexible diaphragm most at mid-span. This, in turn, could move the supports out of their bearing position and result in collapse from gravity loads, a common type of failure.

With a building of slender proportions, it is best to use interior partitions as shear walls (see Figure 42). This could prevent failures caused by the diaphragm deflection as well as reduce the required diaphragm strength. However, it should be noted that the interior partition (as shown) would receive twice the shear load of the exterior end walls. Interior partition shear walls require special design consideration. They must be strong enough to resist the increased shear load, attached securely to the diaphragm and must be connected to the foundation to prevent both sliding and overturning.

Main-Frame Shear Walls

Solid masonry works very well in shear walls. Even marginal quality masonry of an 8-inch solid wall can carry approximately 1,000 plf. There are a number of ways to improve shear strength in masonry to as much as 10 times this value. For example, improved quality of masonry units and mortar, special inspection, increased wall thickness and use of reinforcing all increase shear strength. Reinforcing (horizontal and vertical) is normally mandatory in areas subject to high winds and earthquakes.

In the apartment building example, the east and west walls would have a greater *unit* shear strength capacity (plf) than the north and south walls. Of course, this is because they are bearing walls that carry a superimposed dead load in addition to the weight of the walls.

It must be kept in mind that shear walls are effective only when the lateral loads are applied parallel to the wall. The roof and floor diaphragms transfer their accumulated shear forces to the walls and require proper attachment between each (see Figure 40).

It should be observed for this building example (see Figure 34) that only a small portion of the north and south walls utilize masonry construction functioning as shear walls. Therefore, it would be good wind-design to assign interior partitions for this purpose. This has been done in a more developed solution shown in Figure 58b.

Solid Masonry Wall Weaknesses

Although there is the potential for great shear strength in solid shear walls, there are also weaknesses to note. One weakness in brittle masonry already has been pointed out. When a diaphragm deflects laterally to such an extent that the bearing wall is pushed out, it may crack and/or fail due to the gravity load. (See Figure 41.)

Another weakness occurs when the wall functions as a component and bends vertically between diaphragms when the wind direction is perpendicular to the wall. This is most critical when walls span greater distances between diaphragms. Vertical reinforcing will normally counter this weakness.

Another unique characteristic of masonry walls that must be understood when designing them as shear walls: Masonry walls are considered to be structurally "brittle." They could fail suddenly and completely with little distortion. This characteristic requires

Figure 43 PROPORTIONING SHEAR FORCES TO PIERS OF UNEQUAL STIFFNESS

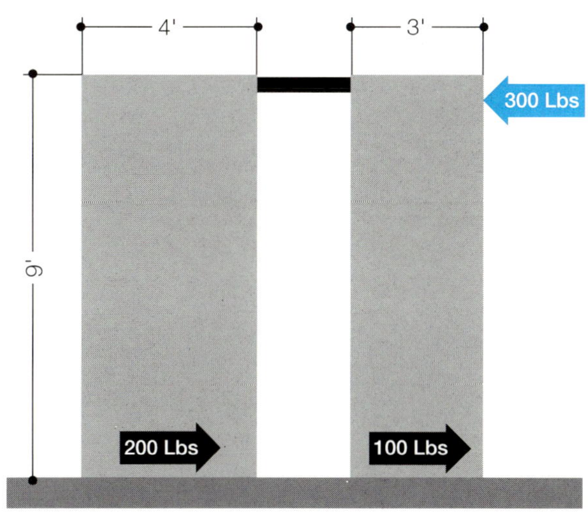

that the shear load be distributed to wall segments according to their relative stiffnesses. When a wall is made up of segments, as occurs when a large wall surface has windows or openings, a series of piers results. If the piers are all the same size (width, thickness and height), the total wall shear load can be divided equally to each segment because they have equal stiffnesses. This condition does not occur in our apartment building example because the exterior walls consist of wall segments of mixed sizes.

If the shear load is distributed equally among masonry walls of different sizes, as illustrated in Figure 43, failure could occur. This is because the stiffer wall will always develop larger forces than the average. The larger force could start a local failure, which could lead to progressive failure of the entire wall.

Figure 43 illustrates two masonry piers of differing widths that are attached to one another at the top. Although one pier is only one-third wider than the other, it has twice the stiffness and will, therefore, develop twice the force. If the total shear force to be resisted is 300 lbs as shown, the four-foot wide pier will develop 200 lbs and the three-foot wide pier will develop just 100 lbs. This procedure for proportioning loads for stiffness is not covered in this monograph, but is not overly complicated. This comparison is simply to emphasize the importance of considering stiffness in brittle materials when forces are distributed to building elements. Overlooking this design requirement could be disastrous. It is equally important to note that if the shear wall is constructed of non-brittle material, such as wood panels attached to studs, this proportioning procedure is not required.

In the apartment building example, the solid masonry exterior walls must therefore be designed using the proportioning procedure. Figure 44 shows the east or west elevation. Although symmetry occurs, there are three different pier sizes. The shear force V_1 will be the largest. The next largest will be V_2, and V_3 will be very small. The numerical sum of all the *Vs* will be equal to all the *Fs*.

Shear Walls–Veneered

As stated earlier, the proportioning procedure would not be required if the exterior walls were constructed of masonry veneer on a structural backing, provided that the backing was non-brittle. Examples of non-bearing and bearing shear walls composed of veneer and structural backing are shown in Figure 45. The veneer does not contribute any strength to the shear wall. The structural backing takes all the shear force and must be designed accordingly. This also applied to its attachment to the diaphragms and to the foundation. In the apartment building example, the foundation anchorage would be more critical for the north and south exterior walls since they are not load-bearing.

Their overturning moment would probably require "tie-downs" as described in the next section.

Figure 44 SOLID MASONRY SHEAR WALLS WITH PIERS OF UNEQUAL STIFFNESS

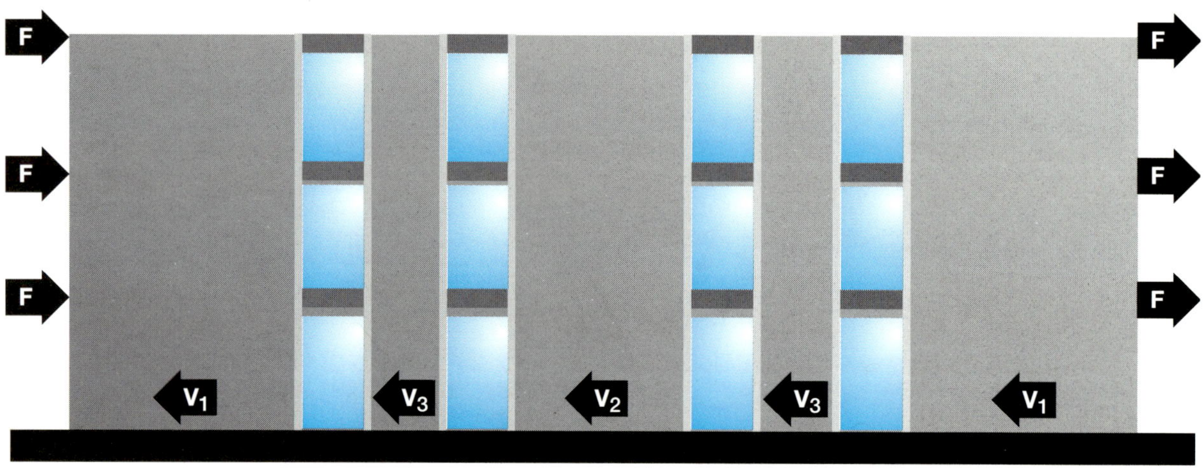

The foundation would have to be sized to counter the overturning effect. The same would be true for solid shear walls.

Interior Partitions as Shear Walls

Interior partitions can also function as effective shear walls. They can be effective whether or not they are used as bearing walls. Their effectiveness reduces the size and deformation of the diaphragm, and they usually develop larger forces than exterior shear walls. Consequently, interior shear wall attachment to the diaphragm and to the floor (foundation) are critical.

Many interior partitions consist of drywall mounted on studs. Placing a wood panel between the drywall and stud is an inexpensive and effective way of achieving a shear wall of meaningful strength. The wood panel can be placed on one side or both sides of the framing. In addition, since this type of shear wall is "flexible," no brittle proportioning is necessary. Its design requires only that the length in any given line be added up and the shear load divided evenly along this length. Safe load tables are available for various construction types and proper detailing of connections and anchorage. Remember, the numerical example presented later uses this apartment building and designates several interior partitions as shear walls.

Example—Single-Span Steel-Framed Building

In this example of a steel-framed building (Figure 35), the wind main-frame in the north-south direction is depicted as a gable-shaped three-hinged arch system. There are five of these arches spaced evenly. Purlins at the eaves and ridge, and in five locations in between, are used to carry the roof gravity load to the arch frames. Roof panels are attached to roof purlins on the frames. Wall panels are attached to wall girts and eaves purlins on the frames. The building would be unstable in the east-west direction without some sort of bracing. As shown in the figure, X-bracing is used in the two center bays. Of course, this example could use other

Figure 45a OVERTURNING VENEER SHEAR WALLS WITH STRUCTURAL BACKING (NON-BEARING)

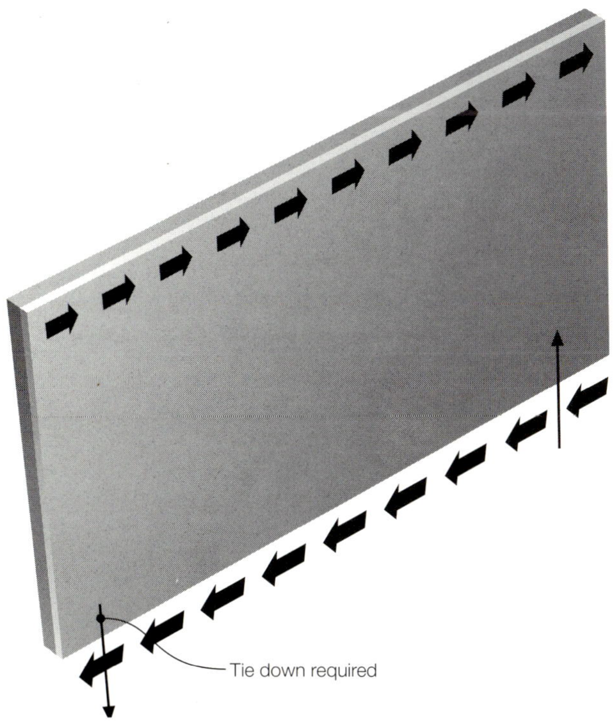

Figure 45b OVERTURNING VENEER SHEAR WALLS WITH STRUCTURAL BACKING (BEARING)

bracing configurations or steel frame types other than the three-hinged arch.

Roof Components

Roof components consist of the panels and the purlins as shown in Figure 46. The panels must be designed for a pressure that acts inward and outward, but not at the same time. They must be strong enough to carry this pressure to the purlins and be attached to the purlins such as to carry the total force on the panel. Care must be taken to account for the increased local wind pressures along the eaves and ridge.

The purlins function as collectors for the wind load and transfer the forces to the frame. This must be in addition to their other task–carrying the gravity loads to the frames. Under these circumstances, when carrying both gravity and wind loads, the allowable stresses can be increased by one-third. If the wind uplift pressure overcomes the gravity dead load carried by the purlins, the connections between purlins and frame must hold the purlins down.

Wall Components

The north and south walls are surfaced with panels which are attached to horizontal wall girts, eaves purlin and a wall sill. The windward side will have the pressure pushing inward (Figure 47a), while the leeward side will have the pressure pulling outward (Figure 47b). Attention must be given to the higher pressures near corners and discontinuous edges. The panels must be strong enough to carry the pressure to their supporting components and the connections must be strong enough for the accumulated pressures, both inward and outward, but not at the same time.

The eaves purlin and wall girts (both are collectors) must be strong enough, in bending capacity, to carry the total surface pressure from the area they support to the frames. Their connections to the frame must be designed for both inward and outward forces.

The two gable ends (east and west surfaces) are particularly critical in this example (see Figure 48). Once again, surface panels are designed in much the

Figure 46 ROOF FRAMING ELEMENTS THAT SERVE AS COMPONENTS AND COLLECTORS

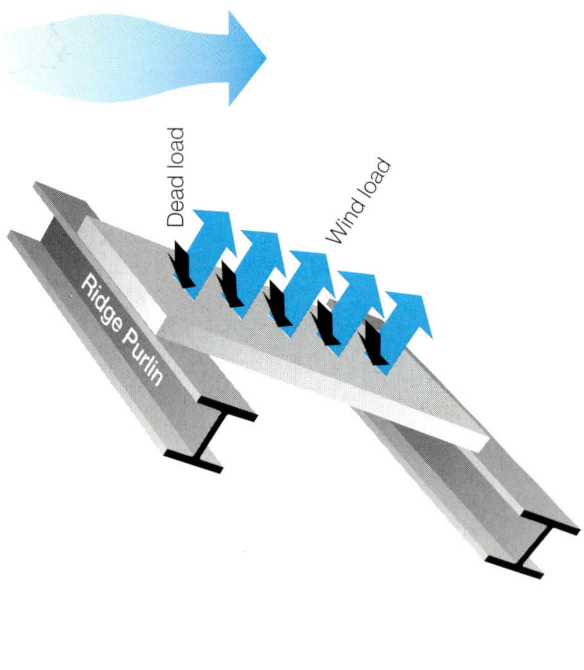

Figure 47a WALL PANEL COMPONENTS (WINDWARD PANEL)

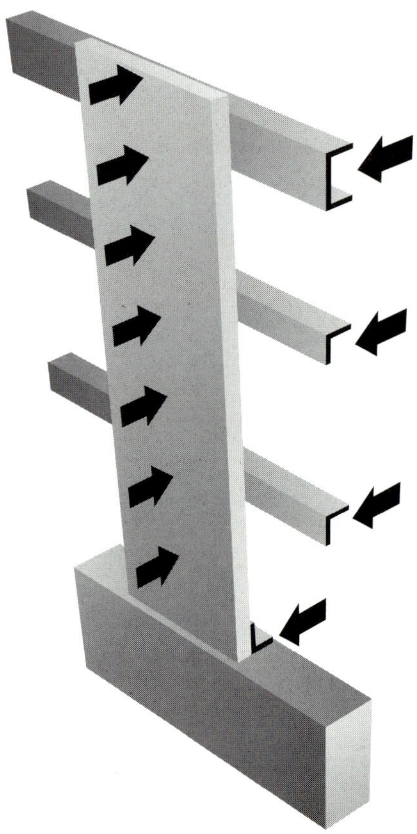

same way as the side walls. However, in this example, they are spanning horizontally. This is because the wall wind collectors span vertically, from the sill to the frame. To function properly, they must be continuous from top to bottom. The horizontal elements serve no structural purpose other than for lateral support of the collectors.

When the collector is long, as near the center of the wall, it will require a significant depth perpendicular to the wall. It would undergo inward bending on the windward side and outward bending on the leeward side. The connections to the frame and sill must be capable of carrying both the inward and the outward forces.

Main-Frame–North-South Direction

Frequently, interior frames are designed to be the same as frames at exterior walls. Since their loads are different (both gravity and wind), each frame must be analyzed independently in order to determine the governing uniform size. Since windward wall forces push in and leeward wall forces pull out, the forces are additive in their overall horizontal effect on the main frames.

The roof forces on the arched-frames could be inward or outward, depending upon the slope of the roof and the relative dimensions of the building. The tables of pressure coefficients in both ASCE 7 references (used later in numerical examples) show an outward force direction for the leeward roof of this example, while the windward roof has both inward and

Figure 47b WALL PANEL COMPONENTS (LEEWARD PANEL)

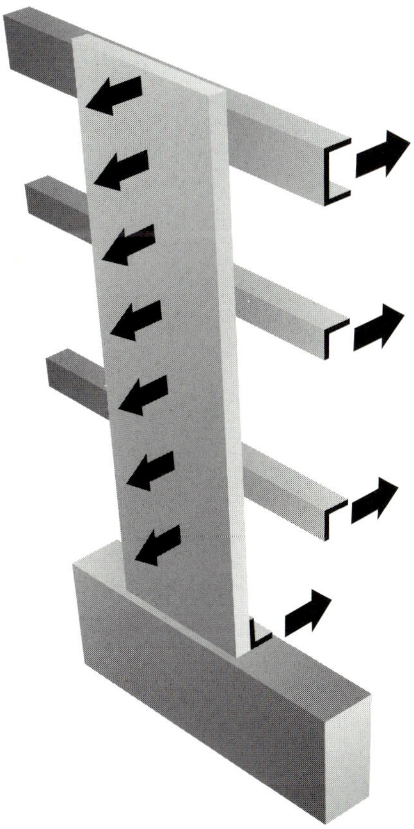

Figure 48 WINDWARD GABLE WALL SHOWING PANELS AND VERTICAL COLLECTORS

outward forces (see Figure 49).

It is necessary to examine the combination of specific forces from each purlin, along with the gravity dead load. The procedure for a three-hinged arch-frame is to first design it for the combined effect of the dead load plus a "balanced" live load. It is then checked and modified, if necessary, for the dead load and an "unbalanced" (one roof side only) live load. Finally, it is checked for the combined dead and wind load, as well as the worst-case scenario of live, dead and wind loads. It should be noted that some codes permit a small reduction in some of the live loads when checking for wind. All codes allow an increase of one-third of the allowable stress when loading includes the short term effect of wind.

An arch-frame requires special attention for its base attachment to the supporting foundation. The gravity loads create an outward thrust at the base of each frame as well as a downward force. Standard anchor bolts are usually designed for this task. However, they must be checked for wind effects. The base on the leeward side would have an additional outward force due to wind effects, for which an equal and opposite reaction force is developed. Finally, the base must be checked for "overturning." The windward side could have a net upward lift, with only the dead load acting to counter the wind effect. A reminder: the dead load stabilizing moment must be 1.5 times the wind overturning moment.

Main-Frame Bracing–East-West Direction

Unlike the three-hinged arch-frames, the bracing does not participate in carrying gravity load. Its sole purpose is stabilizing the building by developing a lateral load-carrying system in the east-west direction. In discussing the wind system in this direction, it is helpful to describe the "load path." The wind pressure is inward on the windward gabled end. The wall collectors carry the forces to the roof line of the end frame. From here the purlins, acting in compression, carry the forces to the bracing bays (Figure 50).

The wind pressure is outward on the leeward gabled end frame. The wall collectors carry the forces to the roof line of the end frame, where they are pulling outward. From there the purlins, acting in tension, carry the forces to the bracing bays. Purlins in the end bays–one in tension and one in compression–are collectors for the main-frame system in this east-west direction. It should now be obvious that these push-pull forces are additive, and that the bracing must be capable of resisting their combined effect. It should also be obvious that these "collector" purlins, whose primary purpose is to carry gravity loads to the frames, must also be designed to carry and transfer the tension and compression forces to the bracing bays.

The roof and wall bracing function together to carry the wind forces to the foundation. The X-bracing implies that the elements are tension-capable only. They are frequently cables or rods, but do not have to

Figure 49a THREE-HINGED ARCH FRAME SHOWING GRAVITY LOADINGS IN CROSS SECTION (LIVE AND DEAD GRAVITY LOADS)

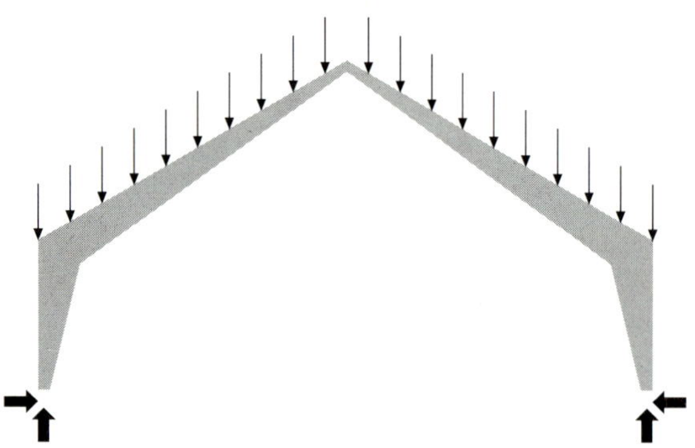

Figure 49b THREE-HINGED ARCH FRAME SHOWING WIND LOADINGS IN CROSS SECTION (WIND LOADS FROM LEFT)

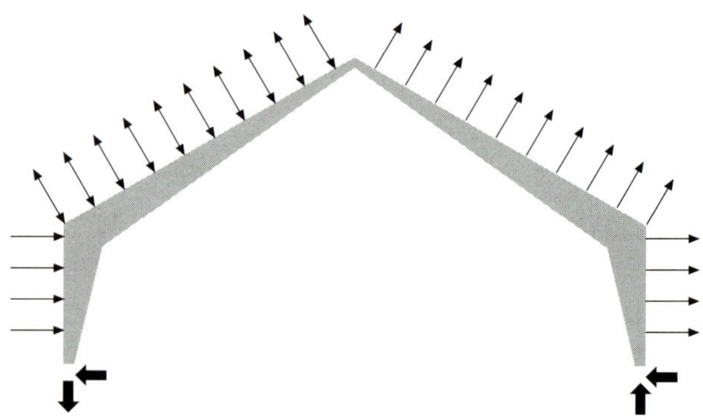

be. The assumption is made that the diagonals are very slender and cannot carry a meaningful compression force. Consequently, only the tension diagonal is effective. Diagonal cross-bracing of this type is shown in Figure 50. The load path reverses in the bracing system (would-be compression diagonals become tension) when wind is active in the opposite direction.

Returning to other design considerations, the purlins in each end bay must be designed to develop the tension and compression forces caused by the gabled end walls. This force is in addition to the bending in the purlin caused by gravity loading. Once more, there is an allowable increase of one-third the allowable stresses. The tension connections are usually the most critical.

Since there are two bracing bays composed of ductile (flexible) material, as opposed to the brittle masonry material used in the previous example, the combined tension and compression forces at the bracing panel points can be divided evenly into the two bracing bays.

They share equally in resisting the wind forces. One of these braced bays is shown in Figure 51 with the "would be" compression diagonal omitted. Wind direction is assumed from east to west. The arch-frame elements will undergo bending about their minor axes when transmitting wind loading from the mid-panel purlins. The mid-panel purlin loads are shown as broken line arrows in the figure. The force system shown is analyzed in a procedure similar to that of a truss. With the diagonals in tension, it should be observed that the purlins at the edges of the braced panels are in compression, the east arch is in compression, and the west arch is in tension. These tension and compression forces in the arches are not usually significant to a degree that would alter their design. The compression in the purlins must be taken into account in their design.

Finally, note the reactions due to the wind. The anchorage at the base of the frames must be designed to withstand the additional forces.

SLIDING - UPLIFT - OVERTURNING

Buildings that are designed and constructed to remain

Figure 50 LOADING ON LATERAL BRACING

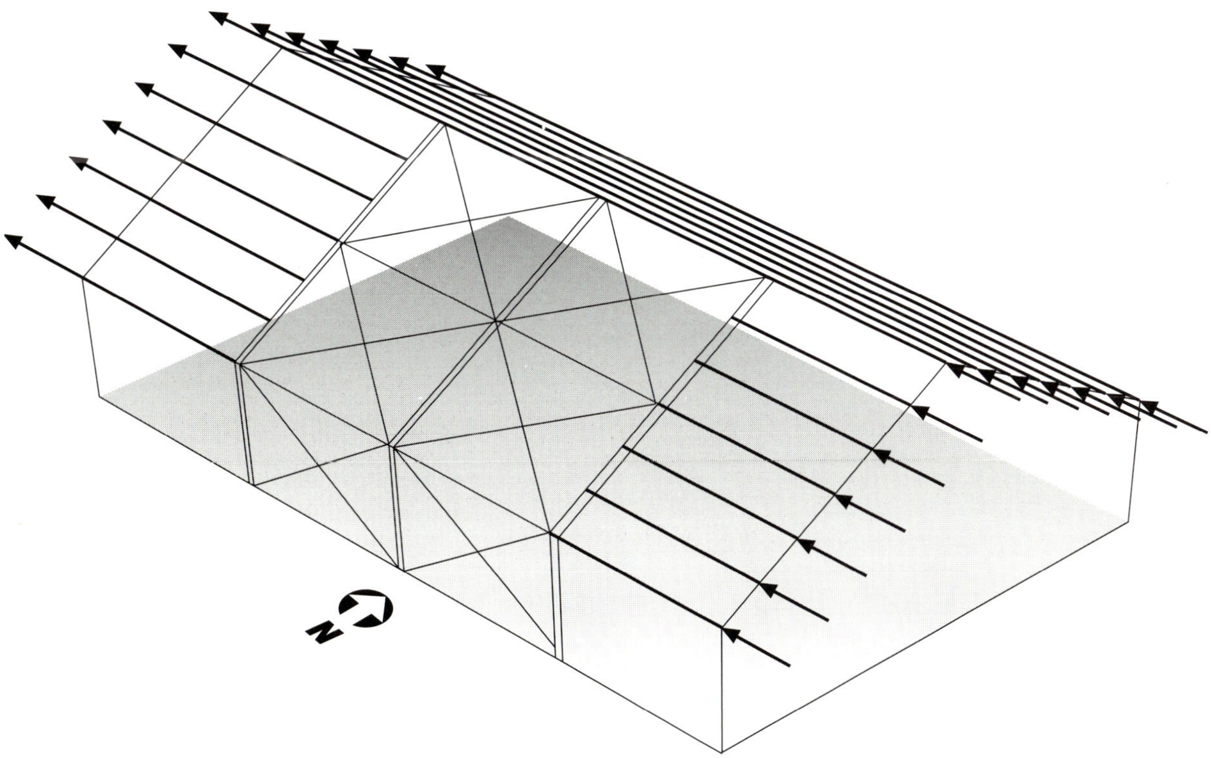

Figure 51 EFFECTIVE BRACING ELEMENTS IN ONE BAY

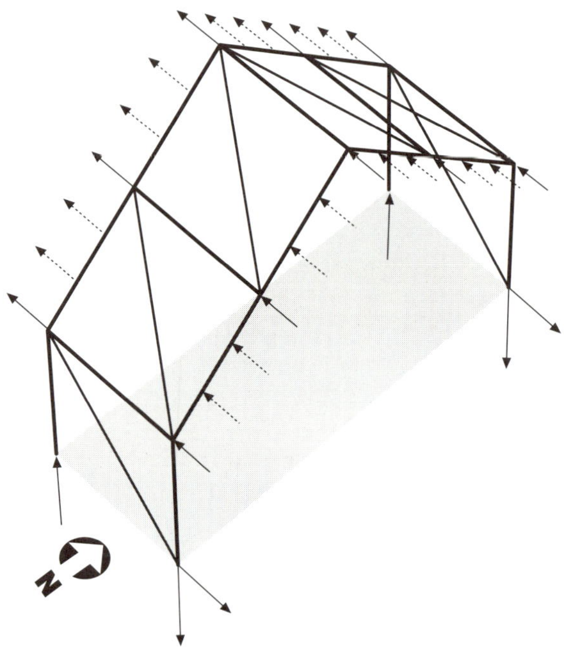

intact can be analyzed as a whole unit and are subject to horizontal sliding, uplift and overturning. (See Figure 52.)

If the foundation of the example building is at or below grade, it will not be affected by the wind. The foundation should be designed not only for gravity loads but also to anchor the building and prevent movement from these other actions. To do this the foundation must be large (heavy) enough to resist the forces from each effect, and the building must be adequately attached to it.

Sliding

Walls designated as shear walls and constructed to function as such are used to resist sliding. Only the designated shear walls need to be bolted to the foundation to prevent the sliding. Engineering calculations indicate that ordinary anchor bolts will be adequate to resist sliding in most cases. For example, one frequently used minimum standard requires one-half-inch anchor bolts, with seven inches of embedment spaced six feet apart to attach a wood sill to a foundation wall. This provision may not be sufficient in all high-wind considerations, but one set of calculations indicates it would be adequate to prevent sliding in a 90-mph

Figure 52a SLIDING ACTION

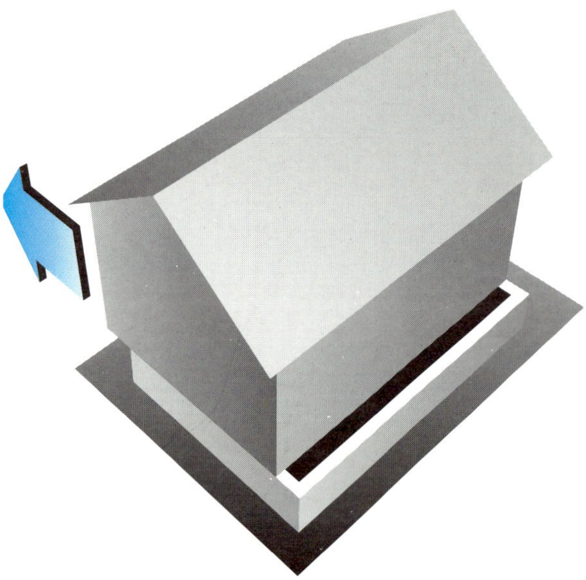

Figure 52b OVERTURNING ACTION

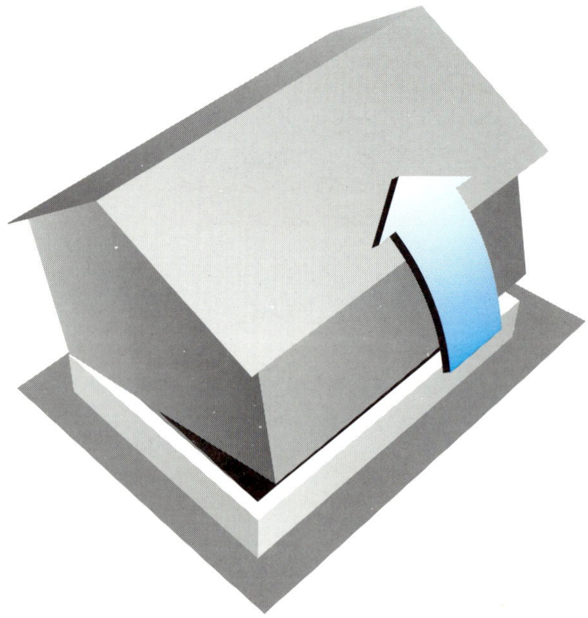

wind acting on a single-story, 25' X 40' building provided that bolts are placed in all exterior walls. Of course, larger bolts and closer spacing should be used when necessary.

Uplift
The weight of a building is usually adequate to resist total uplift. Exceptions to this rule are buildings classified as "open" and "mobile homes." Under these circumstances the full wind can get inside and/or under the building, so measures should be taken to prevent uplift.

Uplift plays an especially important role in the design of components, such as roofing. Roofing is commonly damaged by uplift, and even heavy, precast concrete planks have been known to move from uplift if not properly anchored.

Overturning
Overturning cannot be estimated as easily as horizontal sliding. This is because the weight of the building plays an important part in the stabilizing effect. Nevertheless, if one ignores the building weight and uses the previously described minimum anchorage (one-half-inch bolts at six-foot centers), it would be adequate to resist overturning of the 25' X 40' building in a 110-mph wind.

Element Overturning
From a practical point of view, the overturning effect is usually more critical with individual elements and main-frames than with the whole building. Whether designing individual elements, main-frames or entire buildings, the standard practice is to use only two-thirds of the stabilizing effect to balance the overturning effect. This is accomplished either by multiplying the stabilizing moment by two-thirds or by multiplying the design overturning moment by 1.5. All dead weights contributing to the stabilizing effect can be used, but because of the uncertainties involved, the safety factor of 1.5 seems justified.

When individual structural elements are examined for overturning, the uplift attachment to the foundation can become critical, and simply increasing the bolt size may not be adequate. There are many types of "hold down" hardware available that are much more effective. Since these hold down devices have a larger force capacity, it is often necessary to attach them from the foundation directly to the vertical framing element (stud or column) rather than to the sill. If not securely attached to the framing, the sill could pull away from the wall assembly.

Example of Stabilizing Against Sliding and Overturning
Design to resist the overturning of elements can best be explained and understood through a simple engineering calculation. Consider a shear wall element that is 10 feet high and 15 feet long which must resist a wind force of 3,000 lbs. applied at the top, parallel to the wall, as indicated in Figure 53.

For the sake of simplicity in the example calculation, the shear wall material is not considered. However, it must have a shear strength capacity of:

$v = 3000/15 = 200$ plf

This is not a large unit shear force, and there are available safe load tables showing several types of shear wall assemblies that will meet this strength

Figure 53 EXAMPLE OF SHEAR-WALL OVERTURNING FORCES

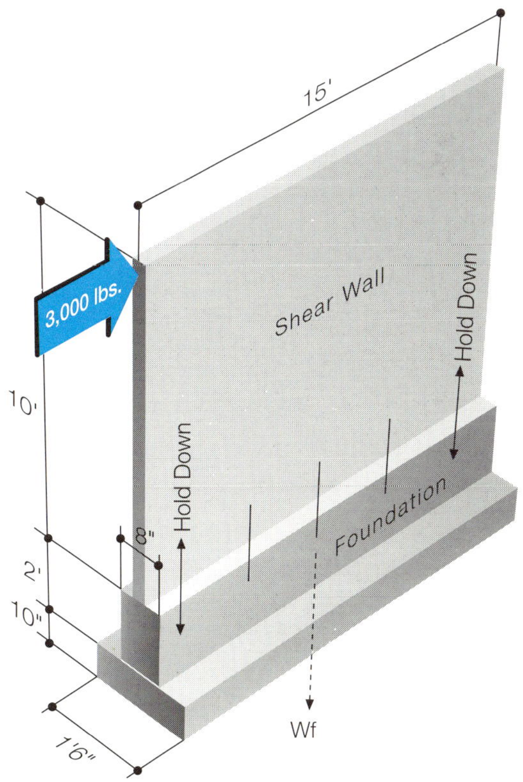

requirement. For larger strength requirements not covered by the tables, the wall assembly would need to be properly engineered and detailed. The overturning moment is designated as *OTM* and, in the example, is due to the lateral load applied at the top of the wall and parallel with the wall. It is calculated for the tendency of the wall to rotate about the lower right hand corner:

OTM = 3000(10) = 30,000 ft-lbs

Also, the total sliding force is seen to be:

V = 3000 lbs

For this example, assume the wall has negligible weight and that there is no friction between the wall and the foundation. Safe load tables will show that, for many cases, the one-half-inch anchor bolt will carry about 675 lbs. in lateral shear. Therefore, the total number of bolts needed to resist the sliding effect is:

Number of bolts = 3000/675 = 4.44

Five bolts will be adequate for the sliding resistance. It should be apparent that the bolts near the center of the wall are not very effective for resisting overturning.

Only the bolts near the ends are considered effective, and for a force acting in only one direction just

Figure 54 HORIZONTALLY POSITIONED WIND COLLECTORS

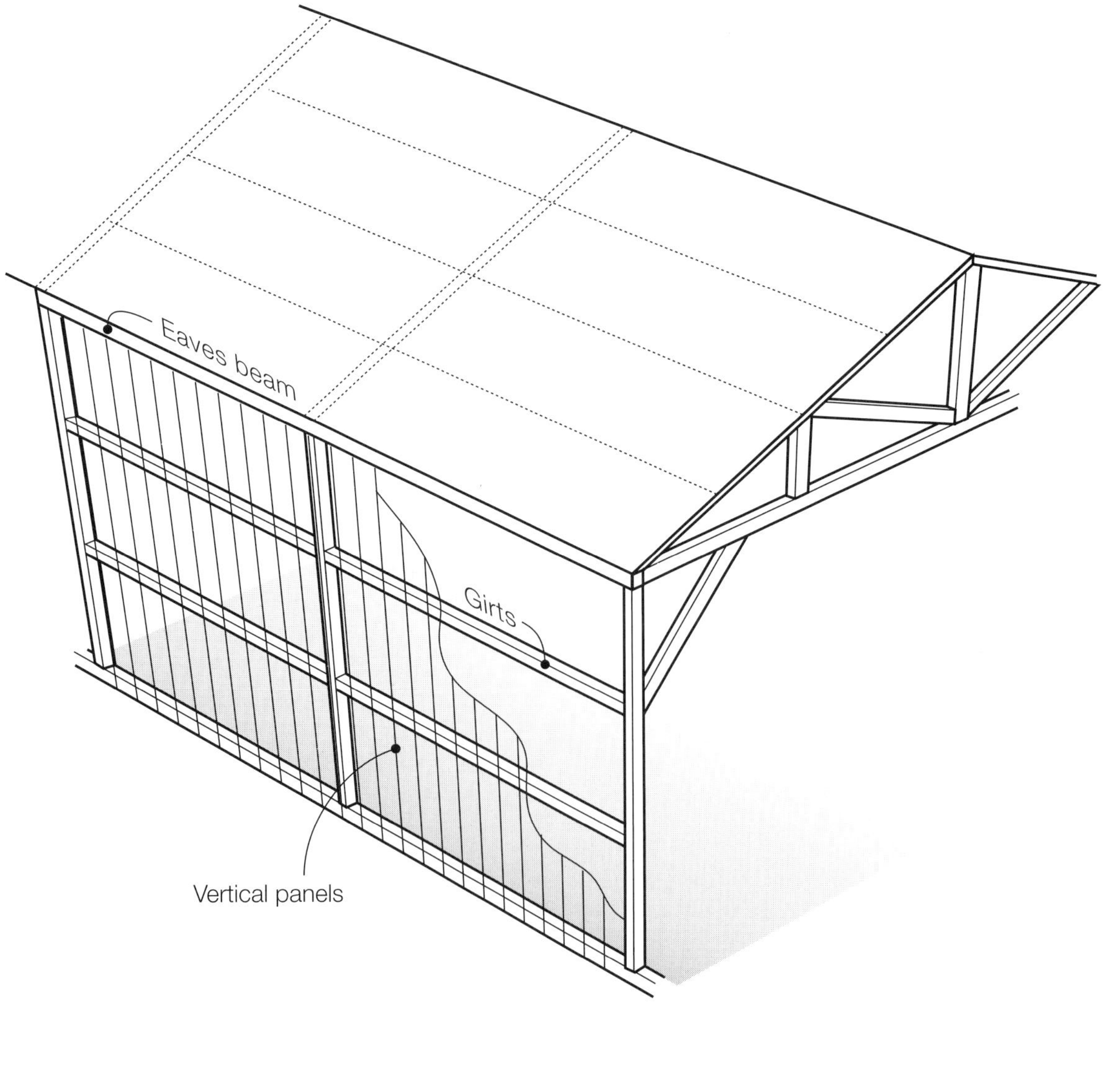

the bolts on the end opposite the point of overturning would be active. Standard practice, of course, considers wind acting in both directions. If the end bolts are located one foot in from each end, their moment arm is 13 feet. They must have a hold-down force *(F)* capacity of:

$(\frac{2}{3})F(13) = OTM = 30{,}000$ ft-lbs, so:

$F = 3460$ lbs

This is a tension force, in the example, for the bolt opposite the point of overturning.

The factor of ⅔ is valid for both the stabilizing moment and its connection. Load tables in manufacturer's catalogues will give anchor size and detail requirements to meet this strength capacity. The two "hold-downs" will also substitute for the anchor bolts that would normally be at the end locations. Observe that the reason for hold-downs at each end of the shear wall is that the shear force (3000 pounds) can occur in either direction.

If the weight of the shear wall is negligible, the only stabilizing weight is that of the foundation. There could be other superimposed weights, such as floors and backfill, but in this case none is shown. The foundation weight (W_f) is seen located at the center of gravity of the foundation wall. The foundation weight required for stabilization is calculated as follows, in which *SM* indicates the stabilizing moment.

$(\frac{2}{3})SM = OTM$

The total weight required to resist the overturning moment, i.e., to produce the required stabilizing moment, is

$\frac{2}{3}(W_f)15/2 = 30{,}000$ ft-lbs, so:

$W_f = 6000$ lbs.

The weight of an 8" X 24" foundation and a 10" X 18" footing, all of concrete, is calculated as follows:

foundation wall = (8/12)(2)(150) = 200 plf

footing = (10)(18)(150)/144 = 188 plf

total weight = 388 plf, so:

Total foundation weight $W_f = 388(15) = 5{,}820$ lbs.

Since 5,820 is less than 6,000 lbs., the wall is inadequate as detailed. A minor modification is needed, such as making the footing width 20 inches rather than the 18 inches as initially assumed.

COLLECTORS

A wind collector functions much like a roof purlin that is receiving a gravity load from a deck and transferring it to a truss or girder. Some collectors receive a wind load from siding and transfer it to the main-frame. Wind collectors function differently than those which are used in seismic design. Most collectors used for seismic design transfer an axial load or a shear load directly to a major member. In wind load effects, most collectors transfer a load by means of bending action. This is a very important observation. A collector, in bending, must be continuous between reactions, otherwise the bending action could not take place.Keep this in mind when considering the following examples.

Horizontal Collectors

Two examples of horizontal wind collectors are wall girts and eaves beams, as shown in Figure 54.

A wall girt usually does not carry much gravity load. It receives the wind pressure from the vertical surface it supports and transfers the load in the form of a reaction to the vertical member into which it is framed.

An eaves beam functions in much the same way, perhaps also with a little gravity load, and picks up the wind load and transfers it at its end connections. If the eaves member were to consist of pieces without a continuous splice, it would be incapable of functioning as a wind collector.

A frequent error is in wall heads (bearing or nonbearing), where the wall's head is supported by studs. If the head were to consist of one member, it would not be long enough and the pieces would simply butt together. This is not a continuous splice, and failure occurs when the eaves member tries to resist bending in the horizontal direction. This is one of the reasons why standard practice is to use double headers that can be properly spliced. The joints can then be staggered and the ends dovetailed around corners.

Observe also that these horizontal collectors gain their strength from depth in a direction perpendicular to the plane of the surface which they are supporting (wall thickness).

Vertical Collectors

Two examples of vertical wind collectors are window mullions and wall studs in a non-bearing wall, as illustrated in Figure 55. Neither of these collectors supports a gravity load. The window glass is certainly non-structural for gravity loads, but it must be structurally capable of sustaining wind pressure. It transfers its surface pressure to mullions which, in turn, span vertically and have a horizontal reaction at the sill and head. The wall studs function in much the same way. In either case, the head may be the edge of a diaphragm of a main-frame, or it could be an eaves beam collector transferring its load, in bending, to a main-frame connection.

Collectors as Components

The collectors described in the preceding sections collect and distribute surface pressures to the main-frame. The surface pressure transferred is that pressure designated for the main-frame. However, it is possible that a collector may also be a component in an area of increased pressure. It should be understood that the component-collector must also be designed for the higher pressure, even though it is not the same pressure delivered to the main-frame.

LATERAL DEFLECTION AND DAMPERS

The dynamic effect of wind on a building causes it to distort and deflect in a very complex manner. Such

Figure 55 VERTICALLY POSITIONED WIND COLLECTORS

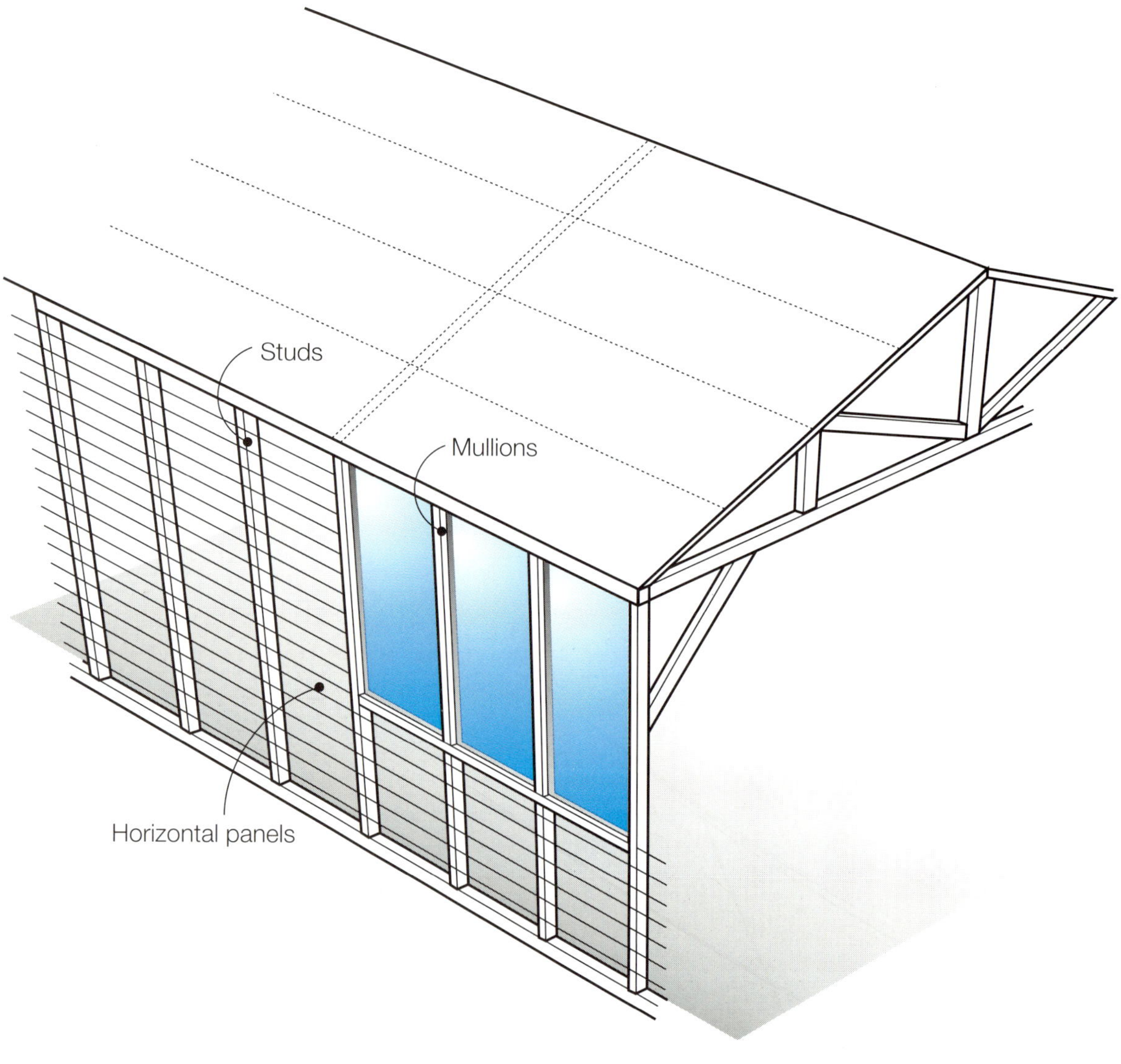

action becomes more critical in tall, slender buildings (skyscrapers), especially if they are of steel construction. Even a mild, steady wind acting on one side of a square or rectangular building can create vortex shedding on the building's sides (oriented parallel to the direction of wind flow). This action tends to move the building back and forth transversely (sideways) to the direction of the wind, a motion which is in addition to the swaying movement in a direction parallel with the wind flow.

Figure 56 VASCOELASTIC DAMPER (TYPE USED IN WORLD TRADE CENTER, NEW YORK)

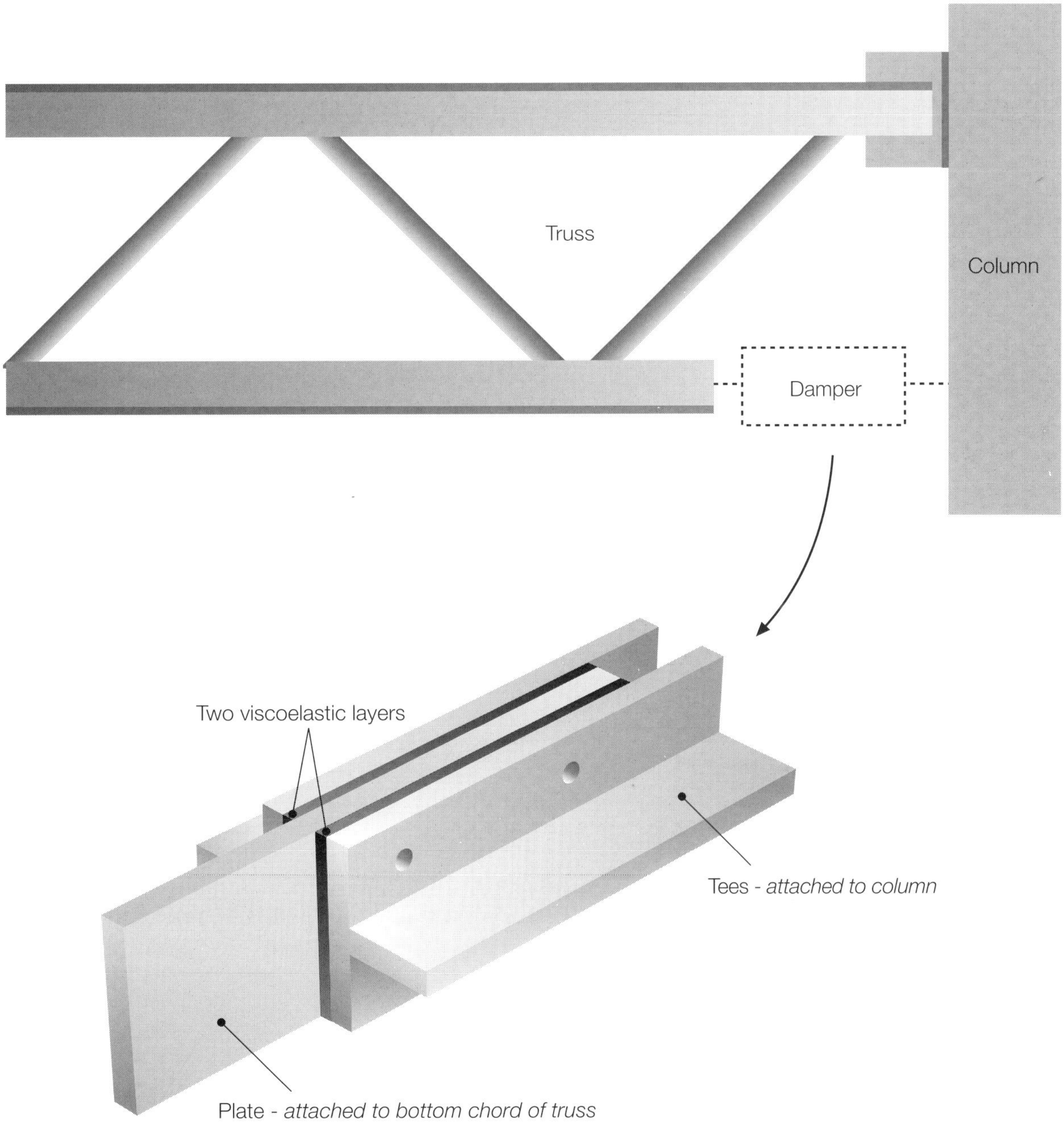

The wind deflection, if excessive, can cause partition cracks, broken windows, and damage to mechanical systems (particularly elevators). Occupants of taller buildings can sense swaying motions and may experience discomfort. Equally important, in modern-day operations, are the effects this movement might have on sensitive equipment found in buildings such as hospitals, recording studios or those housing scientific or computer operations.

There are no prescribed limits for wind deflection in codes and standards at this time. However, a rule commonly used by building designers is to limit the maximum deflection caused by the static design pressures to *H*/500, where *H* is the height of the building. This rule has worked fairly well in keeping lateral deformations within acceptable ranges but might not be adequate for structures that are especially flexible, or for more sensitive operations mentioned above.

There are several ways to reduce the wind deflection of a building. The most common way is to increase the mass weight and/or the structural stiffness of the building. Although it is related, structural stiffness should not be confused with building strength. When mass and stiffness cannot be satisfactorily used to limit a building's lateral movement, mechanical dampers may be an appropriate solution.

Two basic types of mechanical dampers are used in buildings to decrease lateral movement caused by wind pressures. The viscoelastic damper, shown in Figure 56, is a relatively small device that functions much like an automatic door closer. These dampers are used in large numbers and are usually attached to ends of the bottom chords of trusses where they frame into columns. Under lateral loading, these dampers either extend or shorten in length by a very small amount (just a few thousands of an inch), and the absorbed energy is dissipated as heat.

The tuned-mass damper, illustrated conceptually in Figure 57, is a very large element, perhaps weighing several hundred tons, placed near the top of a skyscraper. The damper is mounted on rollers or on a slippery film, which allows the building to move laterally somewhat independently of the damper. The damper's heavy mass tends to remain stationary; as the building moves, spring-like devices at each end extend or compress. The springs tend to pull and push the building back into its undeflected position, thereby reducing the building's overall lateral deformation.

Figure 57 TUNED MASS DAMPER

EXAMPLES OF WIND PRESSURE ANALYSIS

This section discusses wind pressure analysis for the two example building types presented earlier. Two analyses are provided for each example–one following procedures of ASCE 7-93 and one following procedures of ASCE 7-95. The pertinent reference will be needed to follow each example's discussion, because tables and figures from the references are not reproduced here. (The references are not needed to answer questions in the monograph quiz.)

The purpose for these analyses is to describe in detail procedures for determining design wind pressures on building surfaces. The next phase would be to determine forces acting upon the buildings due to these pressures, and then to complete the structural design for wind resistance.

For reason of brevity in the examples (determination of pressures for each surface entails a considerable amount of computation), complete procedures are not included for all building surfaces. For some surfaces, values for coefficients are identified and design pressures are provided without showing all numerical work. Summaries of design pressures, both for main-frames and for components and cladding, are provided in each example.

Example 1—Apartment Building Wind Pressure Analysis Using ASCE 7-93 Procedures

This example uses a three-story, flat-roofed apartment building as shown in Figure 34. A typical apartment plan is shown in Figure 58a.

The main-frame elements are shear walls and diaphragms. Figure 58b shows the vertical elements of the primary lateral-force resisting system in plan view.

Assumptions for analysis:

- Suburban location (Exposure Category B).
- Basic wind speed = 90 mph (value from map based upon fastest-mile wind speed)

Design wind pressures (in psf) will be determined in this analysis. These pressures include those used for design of the main-frame (Part 1) and pressures

Figure 58a EXAMPLE APARTMENT BUILDING FLOOR PLAN

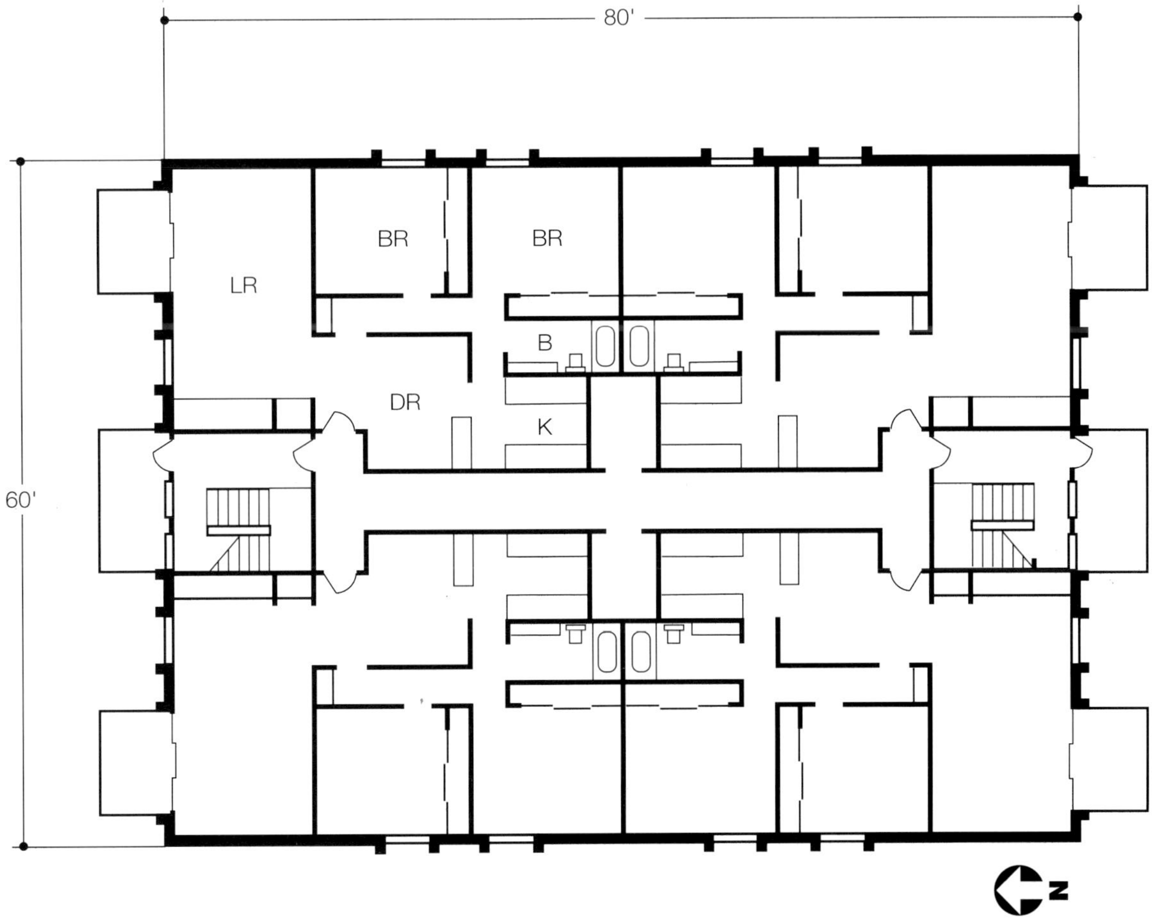

used for design of individual components and cladding (Part 2).

ASCE 7-93, *Minimum Design Loads for Buildings and Other Structures*, is the reference used in this example. It will be needed for those readers who wish to verify specific procedures and numerical values in the example. However, for convenience, the relevant data from the reference are provided as needed for verification of values and computations.

ASCE 7-93 distinguishes between buildings more than 60 feet high and those less than 60 feet high for determination of pressures on components and cladding but not for determination of pressures on the main-frame. Somewhat different procedures are prescribed for components and cladding of buildings less than 60 feet high and those above 60 feet, and different data tables are included for each classification.

Data Having Repeated Use in the Analysis
The analysis of this example is facilitated by first determining the values that have repeated use. Repetitious calculations are minimized if these values are determined at the outset of the analysis.

Velocity Pressure
The equation for velocity pressure q is:

$$q = 0.00256\ K_z\ (IV)^2$$

For this example, the importance factor I is 1.00. Note that if the value for I were greater than 1.00, it would be multiplied by the basic wind speed V before squaring the term. This treatment of I is unique in ASCE 7-93.

Velocity Pressure Exposure Coefficient
The term K_z in the equation converts the basic wind speed to the applicable exposure category (Exposure B) and also for a particular height z above the ground surface. Values for K_z are provided in Table 6 of ASCE 7-93 for the four standard exposure categories and for selected height increments. If step increments are

Figure 58b VERTICAL ELEMENTS OF MAIN FRAME

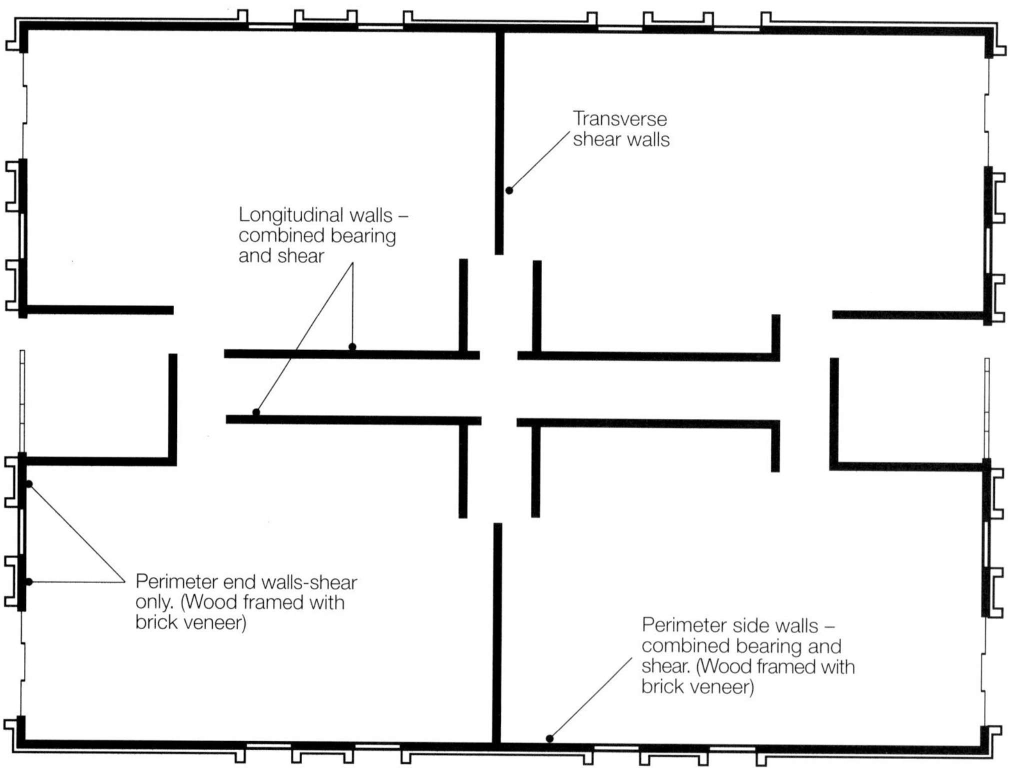

desirable or appropriate due to a particular building's height configuration, interpolated values at other heights are acceptable.

For this example of a 30-foot high building, several values of K_z will be needed. First, a stepped-pressure profile is needed for the windward surface, based upon the step increments of the referenced standard (0 to 15 feet, 20 feet, 25 feet and 30 feet) with a value of K_z needed for each height. Second, leeward, side surfaces and roof surface velocity pressures are based on a constant mean roof height, which in this example is 30 feet, and a value of K_z (typically called K_h for this use, where h denotes mean roof height) at a height of 30 ft is needed. It happens, because the example building has a flat roof, that the value of K_h coincides with the uppermost step increment of the windward surface. Note that with a sloped roof, these two heights would likely be different. Values of K_z for Exposure Category B and for the height increments given above, taken from Table 6 in ASCE 7-93, are as follows.

$K_{z\ 0\text{-}15\,ft} = 0.37$
$K_{z\ 20\,ft} = 0.42$
$K_{z\ 25\,ft} = 0.46$
$K_{z\ 30\,ft} = 0.50$

For convenience, the step increment at 25 feet might be omitted since the floor-to-floor height in this example is 10 feet. By doing so, the pressure step for the third story would coincide with the story height, and therefore simplify the force calculations. In order to provide a complete pressure analysis, we have included the 25-foot height increment.

Part 1. Pressures for Design of the Main-Frame

The general form of the pressure equation, as it appears in Table 4 of the referenced standard is:

$$p = qG_hC_p - q(GC_{pi})$$

Subscripts for the velocity pressure q are omitted in the general equation, because they have a different meaning for windward surface and leeward, side, and roof surfaces. Values for the gust factor G_h will be needed for corresponding heights of K. Table 8 in the referenced standard provides values for G_h for the four standard exposure categories and at various heights. Values for G (called G_z for this use) at corresponding heights for K_z (also K_h when h = 30 ft) are as follows.

$G_{z\ 0\text{-}15\,ft} = 1.65$
$G_{z\ 20\,ft} = 1.59$
$G_{z\ 25\,ft} = 1.54$
$G_{z\ 30\,ft} = 1.51$

It is also convenient to have applicable pressure coefficients C_p for all surfaces at hand. Pressure coefficients for walls and roofs are provided in Figure 2 of ASCE 7-93. In some cases, the applicable coefficient is dependent upon the building's proportions relative to the direction of the wind acting on it, hence the applicable coefficients must be selected carefully.

Pressure coefficients C_p have validity only in context with the direction of wind flow. However, a complete analysis requires examination of wind flow in both directions for each of the two principal axes of the building. Controlling design forces must be determined for each of four possible directions of wind flow. For a symmetrical building, wind pressure analysis can be reduced to a consideration of wind flow in one direction for each of the principal building axes, and pressures already calculated can be correlated for flow in the other directions. For a non-symmetrical building, four pressure analyses may be necessary. For this example, detailed analysis is made for a west-to-east wind flow (see Figure 34). Design pressures are provided for the other three directions of wind flow, but detailed calculations are omitted here.

It is to be noted that the pressure coefficient for the leeward surface of this example is dependent upon a ratio of building length L (length in the direction of wind flow) to building breadth B (width perpendicular to the direction of wind flow). Because the pressure analysis must consider wind on both the long and short building faces of the example building, these ratios will be different for each direction of wind flow being considered; i.e., either 60 ft/80 ft = 0.75, or 80 ft/60 ft = 1.33. The C_p for these two cases are, respectively, -0.5 and -0.43 (interpolated value). For simplicity in the example, -0.5 is used for both cases–a conservative simplification.

The applicable pressure coefficients for all of the building's surfaces are:

Windward surface:	C_p = + 0.8
Leeward surface:	C_p = - 0.5
Side surfaces:	C_p = - 0.7
Roof surface:	C_p = - 0.7

Use of the pressure equation given above also requires values for the combined pressure/gusting coefficient (GC_{pi}) which takes internal effects into account. Values for this combined coefficient are found in Table 9 of ASCE 7-93.

Close reading of the table and its footnotes will reveal that Condition I (the general case rather than an exception specified in Condition II) applies to this example building. Two values of the combined coefficient apply to every building–one value implying a pressure acting toward the surface, the other implying a pressure acting away from the surface. The combined internal pressure coefficients (GC_{pi}) for this example are:

(GC_{pi}) = + 0.25 and - 0.25

We now proceed to determine the pressures on all surfaces for the main-frame of this example. Because velocity pressures at several height increments are needed, these calculations are completed first in order to simplify the design pressure calculations and avoid repetition.

Values for q_z at the relevant height increments are calculated using the equation for q given above.

$q_{0\text{-}15\,ft}$ = 0.00256 (0.37) (1.00 x 90)^2 = 7.67 psf
$q_{20\,ft}$ = 0.00256 (0.42) (1.00 x 90)^2 = 8.71 psf
$q_{25\,ft}$ = 0.00256 (0.46) (1.00 x 90)^2 = 9.54 psf
$q_{30\,ft}$ = 0.00256 (0.50) (1.00 x 90)$_2$ = 10.37 psf

Windward Surface–Wind Direction West to East

Values for a stepped-pressure profile are determined from the following equation.

$p_z = q_z G_z C_p - q_h (GC_{pi})$

Next, calculated pressures at relevant height increments are provided. Note that two design pressures are obtained at each height increment because two directions of internal pressure must be considered.

$p_{0\text{-}15\,ft}$ = 7.67 psf (1.65)(+0.8) - 10.37 psf (+0.25)
= + 10.12 - 2.59
= + 7.53 psf

= 7.67 psf (1.65)(+0.8) - 10.37 psf (-0.25)
= + 10.12 + 2.59
= **+ 12.71 psf**

$p_{20\,ft}$ = 8.71 psf (1.59)(+0.8) - 10.37 psf (+0.25)
= + 11.08 - 2.59
= + 8.49 psf

= 8.71 psf (1.59)(+0.8) - 10.37 psf (-0.25)
= + 11.08 + 2.59
= **+ 13.67 psf**

$p_{25\,ft}$ = 9.54 psf (1.54)(+0.8) - 10.37 psf (+0.25)
= + 11.75 - 2.59
= + 9.16 psf

= 9.54 psf (1.54)(+0.8) - 10.37 psf (-0.25)
= + 11.75 + 2.59
= **+ 14.34 psf**

$p_{30\,ft}$ = 10.37 psf (1.51)(+0.8) - 10.37 psf (+0.25)
= + 12.53 - 2.59
= + 9.94 psf

= 10.37 psf (1.51)(+0.8) - 10.37 psf (-0.25)
= + 12.53 + 2.59
= **+ 15.12 psf**

The controlling design pressures are shown above in bold type. In fact, computation time can be reduced by first observing the algebraic combinations which will produce the controlling pressures. The resulting pressure profile is shown in Figure 59a.

Leeward Surface

This design pressure is constant for the full height of the building and is based on the following equation. It also has two values for internal pressures.

$p = q_h G_h C_p - q_h (GC_{pi})$

Hence:

p = 10.37 psf (1.51)(-0.5) - 10.37 psf (+0.25)
= - 7.83 - 2.59
= **- 10.42 psf**

p = 10.37 psf (1.51)(-0.5) - 10.37 psf (-0.25)
= - 7.83 + 2.59
= - 5.24 psf

Side Surfaces

The design pressure for each side surface is constant for the full height of the building, based on the same equation used for the leeward surface, and it also has two values for internal pressure. The calculated side surface pressures are:

p = 10.37 psf (1.51)(-0.7) - 10.37 psf (+0.25)
= - 10.96 - 2.59
= **- 13.55 psf**

p = 10.37 psf (1.51)(-0.7) - 10.37 psf (-0.25)
= - 10.96 + 2.59
= - 8.37 psf

Roof Surface

The design pressure for the flat roof surface is constant over the entire surface, as well as with two internal pressures.

Calculations are the same as for side surfaces because the same pressure coefficients apply.

p = 10.37 psf (1.51)(-0.7) - 10.37 psf (+0.25)
= - 10.96 - 2.59
= **- 13.55 psf**

p = 10.37 psf (1.51)(-0.7) - 10.37 psf (-0.25)
= - 10.96 + 2.59
= - 8.37 psf

Minimum Design Pressures

A minimum design wind load for the main-frame is prescribed in ASCE 7-93. The minimum wind load may not be less than 10 psf multiplied by the area of the building projected on a vertical plane normal to the wind direction. However, no minimum design pressure for roofs is prescribed in ASCE 7-93. It is observed that the force vector for the algebraic sum of controlling pressures on windward and leeward surfaces for the example building exceeds the 10 psf minimum. Accordingly, the design pressures are established by values calculated above.

Design Wind Pressures for the Main-Frame

Design pressures calculated in preceding paragraphs for the main-frame of this example are shown in Figure 59. Recall that these design pressures are for a wind flowing from west to east. Calculations for other directions of wind flow, although not provided here, are substantially the same (because of our choice to use C_p = -0.5 for both orthogonal directions of wind flow) and will produce the same design values. This may not be true for buildings of other configurations and proportions.

Observe that because pressures on side surfaces (relative to direction of wind flow) cancel each other, these will not affect main-frame forces for each of the directions of wind flow. In this regard, side pressure values are not needed to complete a full analysis of wind flow for the main-frame. Side pressures still must be resisted, however, by the walls. In Part 2 of this analysis, we will find that wall pressures obtained from separate consideration of components and cladding will be larger than side surface pressures obtained in the main-frame analysis. Accordingly, side-surface pressures in the main-frame analysis are more of academic than practical interest. Repetitive analytical tasks can be eliminated as familiarity with procedures improves.

Figure 59a EXAMPLE APARTMENT BUILDING MAIN-FRAME DESIGN WIND PRESSURES (LONGITUDINAL SECTION)

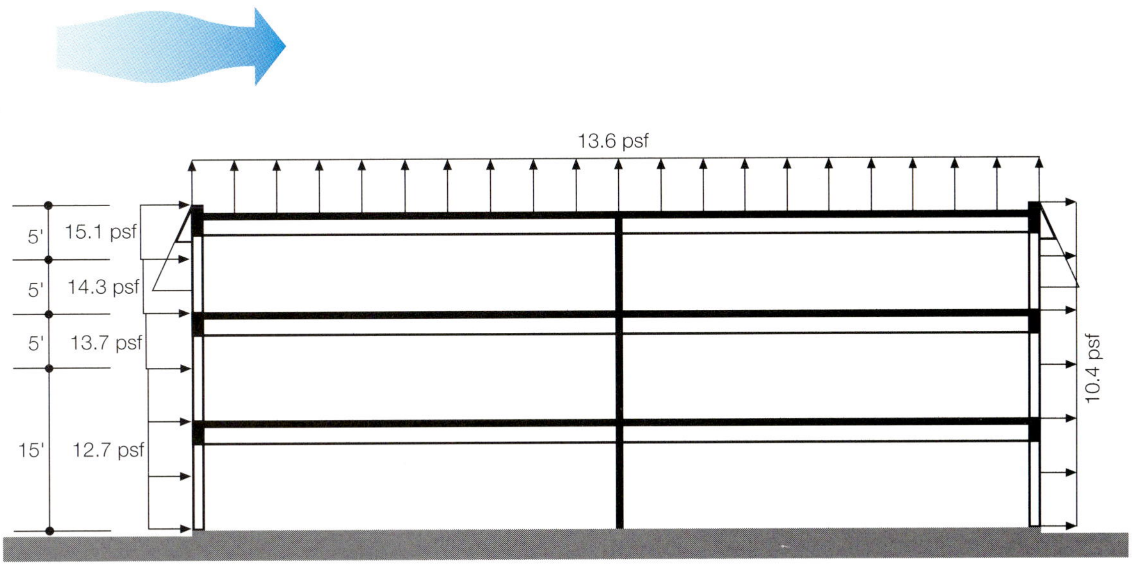

Part 2. Pressures for Design of Components and Cladding

Unlike design pressures for the main-frame, design pressures for components and cladding are independent of the direction of wind flow. All parts of exterior surfaces are considered when determining pressures on components and cladding. Wall and roof surfaces are subdivided into smaller areas for which differing combined pressure/gusting coefficients apply. There are two subdivisions of wall surface areas–corners and inner area, and are three subdivisions of roof surface areas–corners, edges and inside areas. These surfaces are shown in Figure 27.

Forces resulting from pressures on components and cladding are not used in the main-frame analysis, but components and cladding, as well as members serving as collectors must be designed for these forces. The concept of load path is utilized here as pressures must be accumulated in accordance with areas of influence upon the members. As noted at the outset of this example, force development (accumulating pressures as force vectors) is not treated here. As stated earlier, only surface pressures are addressed in this monograph.

In this example, the building is less than 60 feet high and localized surfaces for which increased pressures are considered are prescribed in Figure 3 of ASCE 7-93. In general, wall and roof area subdivisions are defined by a dimension a (see Figure 27). This dimension defines the "strip width" of a wall corner and roof edge, and both side dimensions of a roof corner. Other dimensions of wall edges, wall inner surfaces, roof edges, and roof inner surfaces are simply what is left over after the dimension a is subtracted. The value of a is dependent upon building dimensions, but may not be less than certain specified values. Specifically, a is the lesser value of 0.10 x least building dimension, or $0.4h$, where h is the average roof height, but not less than either 0.04 x least building width, or 3 ft.

The dimension a is needed to determine the area of each subdivided surface for the example building, so it is calculated now for all uses later:

0.1 x least building width = 0.1 (60 ft) = 6 ft

0.4 x h = 0.4 (30 ft) = 12 ft

The lesser value is 6 ft.

0.04 x least building width = 0.04 (60 ft) = 2.4 ft

The 6-foot value is greater than 2.4 ft, so a value of a = 6 ft is used.

The general form of the pressure equation for components and cladding, as it appears in ASCE 7-93, Table 4, and including all pertinent subscripts, is:

$$p = q_h[(GC_p) - (GC_{pi})]$$

Note that pressure coefficients (GC_p) and (GC_{pi}) are added algebraically before multiplying by the velocity pressure. Subscripts included in the general

Figure 59b EXAMPLE APARTMENT BUILDING MAIN-FRAME DESIGN WIND PRESSURES (TRANSVERSE SECTION)

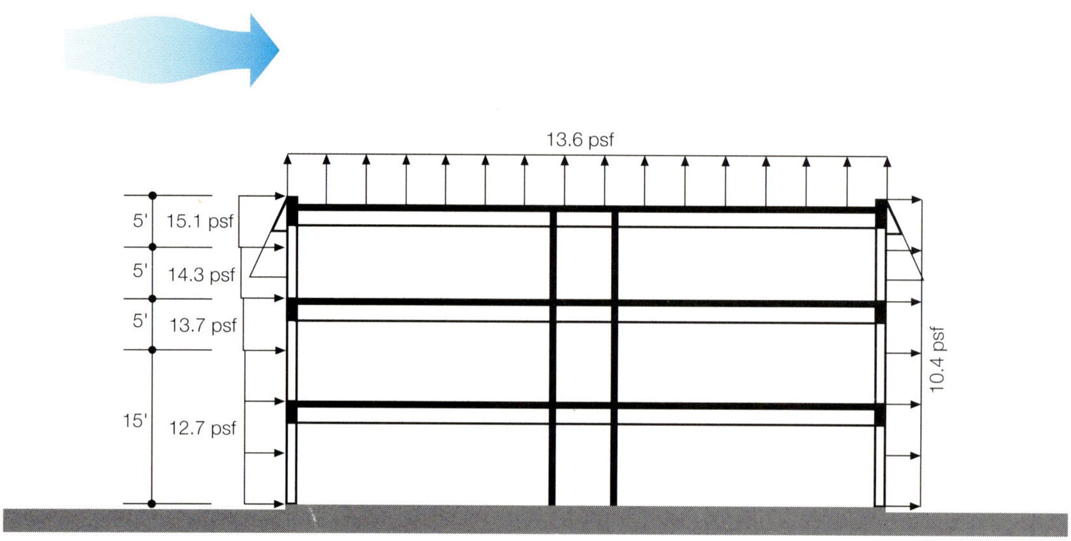

equation require additional comment. The velocity pressure q_h is determined at the mean roof height for Exposure Category C, regardless of the actual building exposure category for buildings with a total height of less than 60 feet. A more complicated form of the equation is used for buildings of greater height, and the actual exposure category is used.

The velocity pressure q_h, based upon Exposure Category C, is the same for all uses of this equation in this example, even though its location is Exposure Category B. The value is computed from the same general equation for q used in Part 1 of the analysis, except for the subscripts. The subscripts h replace the subscripts z in the prior equation and simply denote that an average roof height (30 ft for the example building) is used. The coefficient K_h is obtained from Table 6 in the referenced standard for Exposure Category C at a height of 30 feet:

$$K_{h\ 30ft} = 0.98$$

The calculated value for velocity pressure, then, is:

$$q_{30ft} = 0.00256\ K_{30ft}\ (IV)^2$$
$$= 0.00256\ (0.98)(1.00 \times 90)^2$$
$$= 20.32 \text{ psf}$$

<u>Combined Pressure Coefficients</u>

Values for the combined coefficient (GC_p) are provided in Figure 3 of ASCE 7-93. Note that coefficients are obtained from "nomographs" which are based on the area of the localized surface. This localized area is not the entire wall or roof surface, but only that portion of the subdivided surface being considered. This coefficient, when used for wall surfaces, may be reduced by 10 percent when the roof slope is less than or equal to 10 degrees.

Values for the combined coefficient (GC_{pi}) are obtained from Table 9 of the referenced standard, and in this example, are the same values used earlier for the main-frame pressure equation–that is, the values of +0.25 and -0.25.

Each and every surface of a building is therefore subdivided into localized areas, and design pressures are calculated for each subdivided area. Because many pressures may be calculated for the subdivided areas, once again controlling design pressures are sought.

The wind flow direction is not a consideration in this analysis, so attention is focused only on areas of subdivisions from which combined pressure coefficients are obtained, often by interpolation, from the appropriate diagrams of Figure 3 in ASCE 7-93. Reduced values for walls are indicated in the values of (GC_p) listed below which are multiplied by a factor of 0.9 (the 10 percent reduction mentioned above).

Wall Corners:

West wall Area = 30 ft (6 ft) = 180 sq ft
(GC_p) = + 1.1(0.9) and - 1.4(0.9)

East wall Area = 30 ft (6 ft) = 180 sq ft
(GC_p) = + 1.1(0.9) and - 1.4(0.9)

North wall Area = 30 ft (6 ft) = 180 sq ft
(GC_p) = + 1.1(0.9) and - 1.4(0.9)

South wall Area = 30 ft (6 ft) = 180 sq ft
(GC_p) = + 1.1(0.9) and - 1.4(0.9)

Wall Inner Areas:

West wall Area = 68 ft (30 ft) = 2040 sq ft
(GC_p) = + 1.0(0.9) and - 1.1(0.9)

East wall Area = 68 ft (30 ft) = 2040 sq ft
(GC_p) = + 1.0(0.9) and - 1.1(0.9)

North wall Area = 48 ft (30 ft) = 1440 sq ft
(GC_p) = + 1.0(0.9) and - 1.1(0.9)

South wall Area = 48 ft (30 ft) = 1440 sq ft
(GC_p) = + 1.0(0.9) and - 1.1(0.9)

Roof Corners: ($\theta \leq 10°$)

Area = 6 ft (6 ft) = 36 sq ft
(GC_p) = - 2.7

Roof Edges: ($\theta \leq 10°$)

West edge Area = 6 ft (68 ft) = 408 sq ft
(GC_p) = - 1.5

East edge Area = 6 ft (68 ft) = 408 sq ft
(GC_p) = - 1.5

North edge Area = 6 ft (48 ft) = 288 sq ft
(GC_p) = - 1.5

South edge Area = 6 ft (48 ft) = 288 sq ft
(GC_p) = - 1.5

Roof Inner Area: ($\theta \leq 10°$)

Area = 68 ft (48 ft) = 3264 sq ft
(GC_p) = - 1.2

<u>Localized Design Pressures</u>

Design pressures for components and cladding are calculated using the pressure equation provided above. Four pressures are calculated for each wall surface subdivision, because each of the two pressure coefficients (GC_p) and (GC_{pi}) have two values; two pressures are calculated for each roof subdivision area, because the term (GC_{pi}) has two values. Analysis time can be saved by observing algebraic combinations that produce controlling pressures; although this is not done for the example.

Wall Corners:

West wall p = 20.32 psf [+ 0.99 + 0.25] = + 25.20 psf
p = 20.32 psf [+ 0.99 - 0.25] = + 15.04 psf
p = 20.32 psf [- 1.26 + 0.25] = - 20.52 psf
p = 20.32 psf [- 1.26 - 0.25] = - 30.68 psf

The east wall, north wall and south wall values are the same as the west wall for this example.

Wall Inner Surfaces:

West wall p = 20.32 psf [+ 0.90 + 0.25] = + 23.37 psf
p = 20.32 psf [+ 0.90 - 0.25] = + 13.21 psf
p = 20.32 psf [- 0.99 + 0.25] = - 15.04 psf
p = 20.32 psf [- 0.99 - 0.25] = - 25.20 psf

The east wall, north wall and south wall inside surface values are the same as the west wall for this example.

Roof Corners:

p = 20.32 psf [- 2.7 + 0.25] = - 49.78 psf
p = 20.32 psf [- 2.7 - 0.25] = - 59.94 psf

Roof Edges:

West edge p = 20.32 psf [- 1.5 + 0.25] = - 25.40 psf
p = 20.32 psf [- 1.5 - 0.25] = - 35.56 psf

Figure 60 EXAMPLE APARTMENT BUILDING COMPONENTS AND CLADDING DESIGN PRESSURES (USING ASCE 7-93 PROCEDURES)

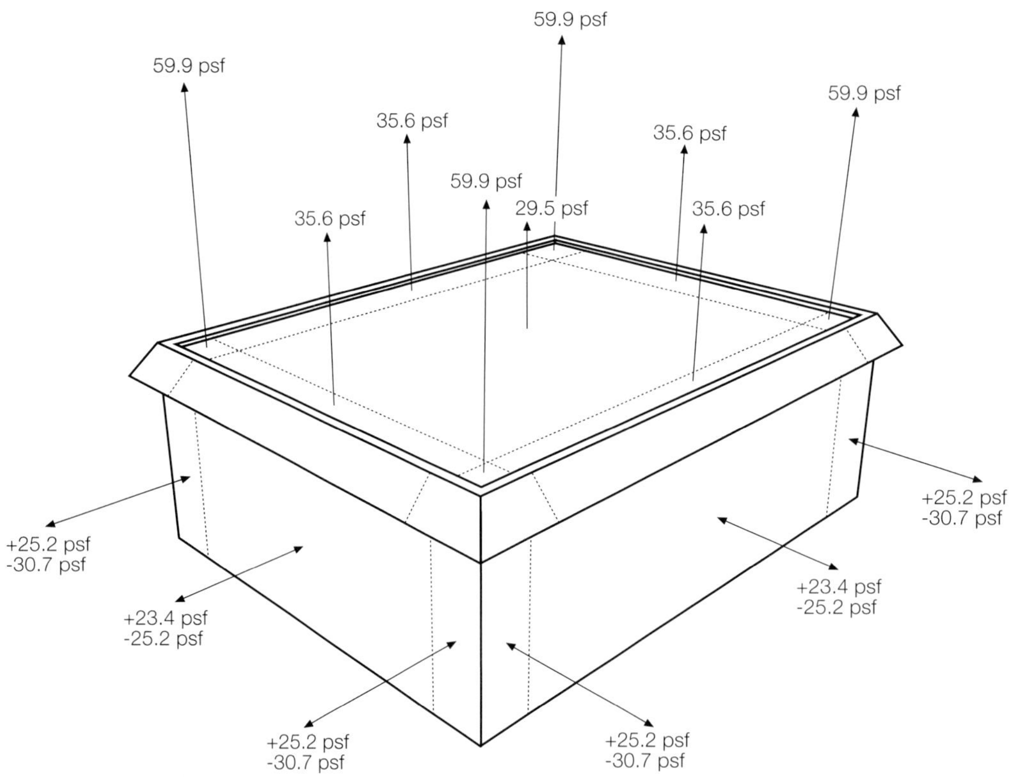

NOTE: Pressures on side surfaces of roof are taken to be the same as corresponding wall surface pressures. These roof surfaces are sloped only slightly from a vertical position, so no significant improvement in accuracy will result from considering them separately.

East edge values are the same as the west edge.
North edge $p = 20.32$ psf [- 1.5 + 0.25] = - 25.40 psf
$p = 20.32$ psf [- 1.5 - 0.25] = - 35.56 psf
South edge values are the same as the north edge.

Roof Inner Area:

$p = 20.32$ psf [- 1.2 + 0.25] = - 19.30 psf
$p = 20.32$ psf [- 1.2 - 0.25] = - 29.46 psf

Design Wind Pressures for Components and Cladding

The component design pressures we have just calculated are shown in Figure 60. Observe that some surfaces have pressures acting both inward and outward. In these cases, elements must be designed to resist them both. The pressures shown are those which apply for all wind conditions when $V = 90$ mph.

Although not provided here, a comprehensive wind analysis would also include determination of pressures on underside surfaces of overhanging roofs and balconies for the example apartment building. Procedures for these calculations are discussed in ASCE 7-93.

Example 2—Apartment Building Wind Pressure Analysis Using ASCE 7-95 Procedures

This example continues to use the three-story, flat-roofed building shown in Figures 34 and 58 used in the ASCE 7-93 example.

Assumptions for analysis:

- Suburban location (Exposure Category B)
- Importance Factor $I = 1.0$
- Basic wind speed = 90 mph (value from map based upon three-second gust wind speed).

Note that a true comparative analysis of ASCE 7-93 and ASCE 7-95 procedures would, in nearly all instances, use a larger basic wind speed in the ASCE 7-95 analysis than that used in ASCE 7-93. This is because velocity contours in the three-second gust wind-speed map generally are larger for most (but not all) of the same locations. However, the increased velocities in the three-second gust wind-speed map do not correlate with velocity contours of the fastest-mile wind speed map, so a direct comparison can be made only for a specific location. Note, for example, that a 90-mph velocity contour in the fastest-mile wind speed map corresponds in all but a few cases with a velocity contour of 110 mph or 120 mph (coastal hurricane region) on the three-second gust wind-speed map.

This example, using ASCE 7-95 procedures, provides a comparison with the same basic wind speed of 90 mph. For most locations, basic wind speeds from the fastest-mile wind map and three-second gust wind map are not the same.

Design wind pressures for the example building are determined in this analysis. These include pressures for design of the main-frame (Part 1) and again, pressures used for the design of the individual components and cladding (Part 2).

ASCE 7-95, Minimum Design Loads for Buildings and Other Structures, is the source for both the procedures and data used in this example. Reference to it is necessary to verify specific procedures and numerical values in the example, but the tables and figure numbers are provided as needed for the computations.

ASCE 7-95 distinguishes between buildings over 60 feet in height and those of lesser height (see Section 6.2 in ASCE 7-95 for qualifications pertaining to "low-rise buildings") for both the main-frame and components and cladding. Slightly different procedures are prescribed for these two height classifications, and different data tables are used for each.

The wind pressure analysis for this example, which has a height of 30 feet, is therefore based on the ASCE 7-95 prescribed procedure for buildings up to 60 feet high. This example building meets the qualifications referenced in the preceding paragraph, since the aspect ratio $h/L = 30/60 < 1.0$.

The values that will have a repeated use are determined first. Repetition can again be minimized if these values are determined at the outset of the analysis.

Velocity Pressure

The applicable equation for velocity pressure q is:

$$q_z = 0.00256\ K_z\ K_{zt}\ V^2\ I$$

This equation is different than the velocity pressure equation of ASCE 7-93. A new term, K_{zt}, is added, and the Importance Factor I is no longer combined with velocity V before squaring. As well, values of I in general are different than in ASCE 7-93.

For this example building, which qualifies as a "low-rise building," ASCE 7-95 prescribes an alternative procedure that is quite different than would otherwise be used. By this alternate procedure, wind actions and consequent pressures on building surfaces are analyzed corner by corner, and surfaces are defined differently than those traditionally done for main-frame analysis. These significant changes in procedures will become evident as the example analysis proceeds. For buildings with heights greater than 60 feet, procedures of ASCE 7-95 are much the same as for ASCE 7-93, although the tabled data are different.

Exposure Category C is used for all "low-rise buildings," regardless of the actual exposure category that is based on location, and a reduction factor of 0.85 is applied to design pressures for buildings sited in Exposure Category B (see footnotes for Table 6-4 and Figure 6-4 in ASCE 7-95). Table 6-1 shows that the velocity pressure for the main frame is based on a value of $z = h$–the mean roof height which for this example is 30 feet. The equation for velocity pressure therefore becomes:

$$q_h = 0.00256\ K_h\ K_{zt}\ V^2\ I$$

Observe a significant result of this subtle change in subscripts: The velocity pressure q has a constant value for the full height of the building (60 ft or less), consequently, there is no increase in velocity pressure as the height increases, and therefore no stepped-pressure profile occurs on the windward surface for "low-rise buildings."

Velocity Pressure Exposure Coefficient

In general, the term K_z in the first velocity pressure equation converts the basic wind speed to the applicable exposure category and also for a particular height z above the ground surface. Values for K_z are provided in Table 6-3 of ASCE 7-95 for the four standard exposure categories and for selected height increments. If different increments are desirable, due to a particular building's height configuration, interpolated values at other heights are acceptable.

Since this example building is 30 feet high, only one value of K_z will be needed ($z = h$, where h is the average height of the roof above the ground surface) for pressure calculations on all building surfaces. The value of K_h for a height of 30 feet for Exposure Category C, taken from Table 6-3 in ASCE 7-95, is:

$$K_{h\ 30\ \text{ft}} = 0.98$$

Topographic Factor

The topographic factor K_{zt} is a new term in ASCE 7-95 and does not appear in ASCE 7-93. It accounts for increased wind velocities associated with hills and other terrain irregularities in near proximity to a building and in particular, buildings sited on the upper half of hills and ridges or near the edges of escarpments. The general equation for this term is:

$$K_{zt} = (1 + K_1 K_2 K_3)^2$$

Values for each of the terms K_1, K_2 and K_3 are provided in Figure 6-2 in ASCE 7-95. Section 6.5.5 provides conditions under which the topographic factor need not be considered. The topographic factor has been ignored in this example analysis because terrain features are not given.

Calculated Velocity Pressure

The working equation for velocity pressure q in this example reduces to:

$$q_h = 0.00256\ K_h\ V^2\ I$$

and the calculated value is:

$$q_h = 0.00256\ (0.98)\ (90)^2\ (1.0) = 20.32\ \text{psf}$$

Part 1. Pressures for Design of the Main-Frame

The pressure equation for "low-rise buildings," as it appears in ASCE 7-95 and including all subscripts, is:

$$p = q_h[(GC_{pf}) - (GC_{pi})]$$

Subscripts are particularly significant in this equation; they have specific meaning and there are specific sources from which their values are obtained. Note that combined gusting/pressure coefficients apply to both the external and internal pressure portions of the equation. However, subscripts of the pressure coefficient C are different in the equation, which implies that values also are different. Indeed, values for each of the combined coefficients (GC_{pf}) and (GC_{pi}) must be separately obtained from different tables.

External Combined Gusting/Pressure Coefficient (GC_{pf})

Values for this combined term are obtained from Figure 6-4, which is new in ASCE 7-95 and also has a different figure/table format than has typically been found in earlier wind codes. Moreover, Figure 6-4 has a number of footnotes which must be read with care.

The first difference encountered in Figure 6-4 is that external pressure is considered in three dimensions of a building–unlike the typical two-dimensional approach of earlier wind codes. A second difference is that two separate cases of wind loading (Case A and Case B) are to be examined for each building to determine the main-frame loading–essentially resulting in a possibility of asymmetrical loading of the main-frame. In effect, the two cases treat wind loading in two half-quadrant regions at a corner, with the procedure to be repeated for each 90-degree quadrant. A third unique aspect of Figure 6-4 is that building surfaces are subdivided–wall and roof surfaces are each divided into two areas, which are considered separately in the analysis. Each subdivided area has a combined external pressure coefficient (GC_{pf}). Since there are eight surfaces for Case A and 12 surfaces for Case B (each with two values for (GC_{pi})), and theoretically, four 90-degree quadrants, a possible total of 160 separate surface pressures may require examination to determine the most severe loading conditions for a building. Clearly, computation efficiency must be sought. A final observation in Figure 6-4 is that the procedure itself, by considering the four corners separately, encompasses full evaluation for wind from any direction acting upon a building.

Internal Combined Gusting/Pressure Coefficient (GC_{pi})
The internal pressure portion of the general pressure equation of ASCE 7-95 combines both the gust effect and the pressure coefficient into a single coefficient (GC_{pi}). Values for this term are provided in Table 6-4 of ASCE 7-95, but these values are not the same as in ASCE 7-93. Reference to Section 6.2 of ASCE 7-95 shows that this example building qualifies as neither an "open building" nor a "partially enclosed building." Accordingly, the values for (GC_{pi}) from Table 6-4 that apply to the example apartment building are:

$$(GC_{pi}) = +0.18 \text{ and } -0.18$$

Pressures for Main-Frame Loading
Using the ASCE 7-95 procedure, determination of the main-frame loading is not simple, although the calculations involve only multiplication. Numerous surfaces occur, each with a unique pressure, even for a simple building shape. Also, directions of the pressure vectors due to external and internal pressure effects must be tracked. Values of all pressures on surfaces are needed

Table 6 APARTMENT BUILDING EXAMPLE–COEFFICIENTS AND CALCULATED PRESSURES FOR MAIN-FRAME CORNER I, CASE A, BASED UPON ASCE 7-95 PROCEDURES

(GC_{pf}) from Figure 6-4, ASCE 7-95	**(GC_{pi})** from Table 6-4, ASCE 7-95	**Building Surface** Figure 6-4, ASCE 7-95	**[(GC_{pf}) - (GC_{pi})] x 0.85**	**Pressure** Column 4 x q (q=20.32)
+0.40	-0.18	1	+0.49	**+10.02**
	+0.18		+0.19	+3.80
-0.69	-0.18	2	-0.43	-8.82
	+0.18		-0.74	**-15.04**
-0.37	-0.18	3	-0.16	-3.29
	+0.18		-0.47	**-9.50**
-0.29	-0.18	4	+0.09	-1.91
	+0.18		+0.40	**-8.13**
+0.61	-0.18	1E	+0.67	**+13.66**
	+0.18		+0.37	+7.44
-1.07	-0.18	2E	-0.76	-15.38
	+0.18		-1.06	**-21.60**
-0.53	-0.18	3E	-0.30	-6.06
	+0.18		-0.60	**-12.26**
-0.43	-0.18	4E	-0.21	-4.33
	+0.18		-0.52	**-10.55**

in order to identify design loadings, but even this information is insufficient to fully understand the main-frame loading. Pressures must be transformed to force resultants on the surfaces, because loading of the main-frame is a function of pressure times surface area, and both typically are different for each surface subdivision in this procedure.

When applied in practice, this lengthy procedure can be shortened through intuitive knowledge of controlling conditions.

However, such intuition evolves only from accomplishing a few complete analyses. For our purposes, the complete analysis is provided for this example. A table format is used to compact the calculations and information. Two tables are used for this example—one for Case A wind loading conditions and another for Case B—in examining the wind actions on just one corner. In order to determine design pressures, at least

Table 7 APARTMENT BUILDING EXAMPLE–COEFFICIENTS AND CALCULATED PRESSURES FOR MAIN-FRAME CORNER I, CASE B, BASED UPON ASCE 7-95 PROCEDURES

Building Surface Figure 6-4, ASCE 7-95	**(GC_{pf})** from Figure 6-4, ASCE 7-95	**(GC_{pi})** from Table 6-4, ASCE 7-95	**[(GC_{pf}) - (GC_{pi})] x 0.85**	**Pressure** Column 4 x q (q=20.32)
1	-0.45	-0.18	-0.23	-4.67
		+0.18	-0.54	**-10.89**
2	-0.69	-0.18	-0.43	-8.82
		+0.18	-0.74	**-15.04**
3	-0.37	-0.18	-0.16	-3.29
		+0.18	-0.47	**-9.51**
4	-0.45	-0.18	-0.23	-4.67
		+0.18	-0.54	**-10.89**
5	+0.40	-0.18	+0.49	**+10.02**
		+0.18	+0.19	+3.80
6	-0.29	-0.18	-0.09	-1.91
		+0.18	-0.40	**-8.13**
1E	-0.48	-0.18	-0.26	-5.18
		+0.18	-0.56	**-11.40**
2E	-1.07	-0.18	-0.76	-15.38
		+0.18	-1.06	**-21.60**
3E	-0.53	-0.18	-0.30	-6.06
		+0.18	-0.60	**-12.27**
4E	-0.48	-0.18	-0.26	-5.18
		+0.18	-0.56	**-11.40**
5E	+ 0.61	-0.18	+0.67	**+13.66**
		+0.18	+0.37	+7.44
6E	-0.43	-0.18	+0.21	-4.33
		+0.18	-0.52	**-10.55**

Figure 61a MAIN FRAME PRESSURES–CORNER I (N.W.), CASE A (USING ASCE 7-95 PROCEDURES)

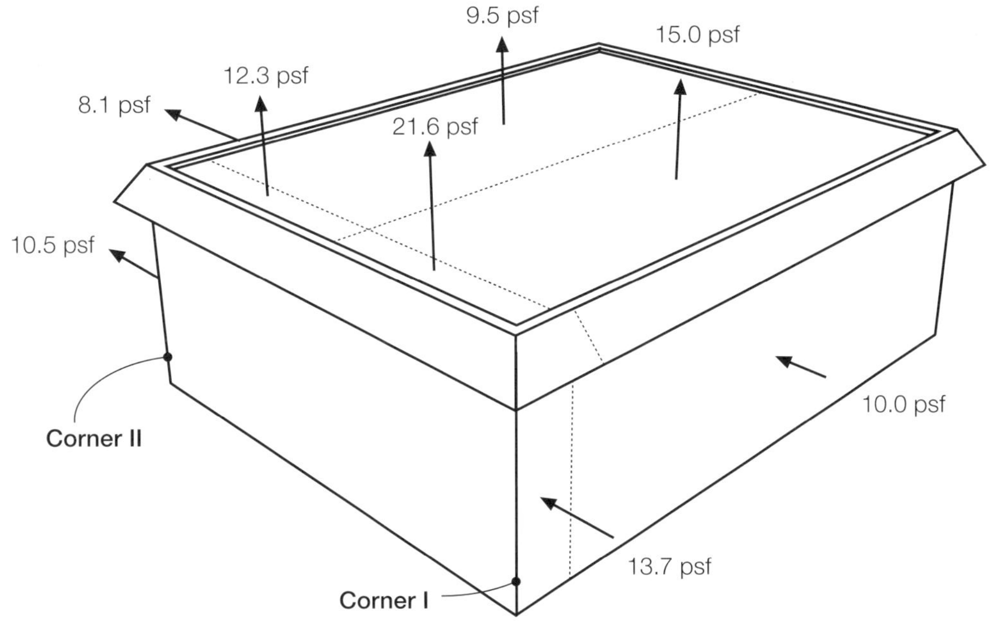

Figure 61b MAIN FRAME PRESSURES–CORNER I (N.W.), CASE B (USING ASCE 7-95 PROCEDURES)

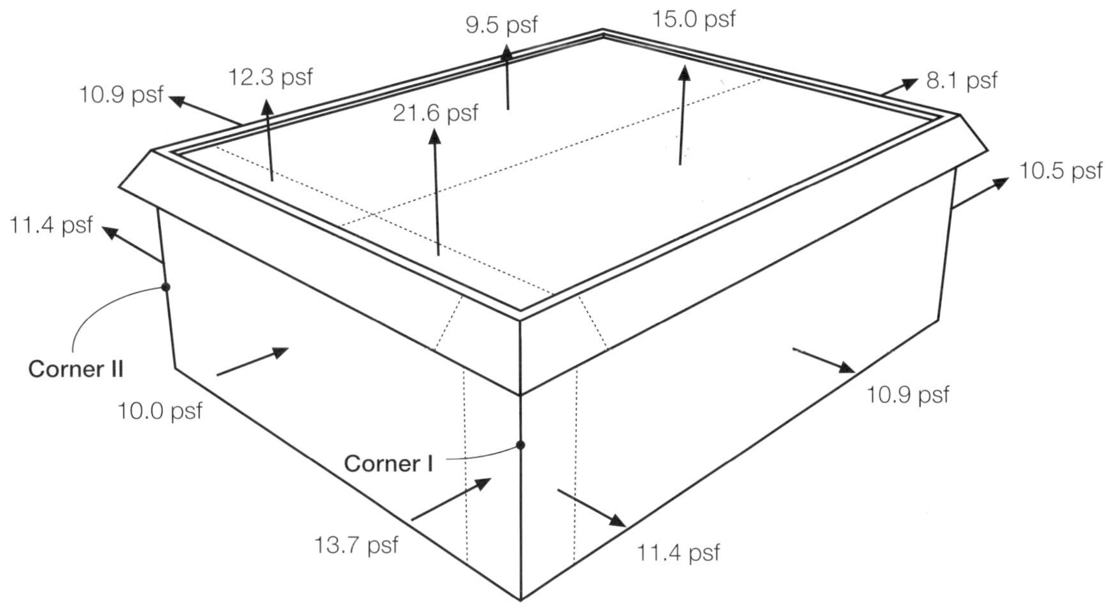

two adjacent corners of the four corners of a symmetrical building require consideration. Pressures for both cases and for both corners of the example building are based on the equation above, which has been modified to include the multiplier of 0.85 (mentioned earlier for buildings in Exposure Category B), as follows:

$$p = q_h[(GC_{pf}) - (GC_{pi})][0.85]$$

Corner I (see Figure 61 for designation of corner and surfaces)

Pressures–Case A. Controlling design pressures for this case are shown in bold type in Table 6.

Pressures–Case B. Controlling design pressures for this case are shown in bold type in Table 7.

Corner II (see Figure 62 for designation of corner and surfaces)

Pressure calculations for Cases A and B for Corner II are the same as for Corner I of the example building, but surfaces must be correlated with Corner II. Such saving of computation time is possible here only because the example building's roof is flat, in contrast with Example 4, where the example building has a gabled roof and pressure calculations for both corner conditions are necessary.

Interpretation of Results

Figure 61 and Figure 62 show the design pressures on the exterior surfaces of the building. There are four separate complex loading patterns to be considered. These surfaces transfer their pressures to the collectors which, in turn, transfer their forces to the main-frame diaphragms and shear walls. It should be apparent that torsional effects on the building are involved. The main-frame must be designed for the most critical condition. This would require four separate force analyses. If computer programs are not readily available for this task, most designers will undertake an overall conservative simplification procedure and use forces larger than the detailed pressure surfaces but with fewer variations.

Section 6.4.1.2 of ASCE 7-95 prescribes a minimum wind load for the design of the main-frame of 10 psf multiplied by the area of the building projected on a plane positioned perpendicular to the direction of the wind. For the example building, the minimum load in each orthogonal direction (calculated as force vectors) is less than the force due to pressures calculated for the main-frame. Force vector calculations are not provided here but are needed to determine the controlling pressures.

Part 2: Pressures for Design of Components and Cladding

Localized surfaces for which increased pressures are considered for this example are prescribed in Figure 6-5 of ASCE 7-95. The figure applies to buildings with a mean roof height equal to or less than 60 feet. Another table is used for buildings over 60 feet in height. All areas of all surfaces are included when considering components and cladding. Wall and roof surfaces are subdivided into smaller areas for which different combined pressure/gusting coefficients apply. There are two subdivisions that apply to wall surfaces (corners and inner areas), and there are three subdivisions that apply to roof surfaces (corners, edges and inner areas). Forces resulting from the pressures on components and cladding are not used in the main-frame analysis, but remember that components and cladding, as well as all members of each surface which serve as collectors of the resulting forces, must be designed for these forces.

Design pressures for components and cladding are independent of the direction of wind flow. Each surface of a building is subdivided into appropriate localized areas, and design pressures are calculated for each subdivision. Because more than one pressure may be calculated for subdivided areas, once again the controlling design pressures are sought.

In general, wall and roof surface area subdivisions are defined by a dimension *a*. This dimension defines the strip width of a wall corner and roof edge, and both side dimensions of a roof corner. Other dimensions of wall corners, wall inner surfaces, roof edges, and roof inner surfaces are simply what is left over after the dimension *a* is subtracted. The value of *a* is dependent upon building dimensions, but may not be less than certain specified values. Determination of *a* follows identical procedures of ASCE 7-93 provided in Example 1 and is not repeated here. The dimension *a* is needed to determine the area of each subdivided surface for the example building, so it is established now for all uses later.

The controlling condition for *a* is 0.1 x least building width = 0.1 (60 ft) = 6 ft

The general form of the pressure equation for

Figure 62a MAIN FRAME PRESSURES–CORNER II (N.E.), CASE A (USING ASCE 7-95 PROCEDURES)

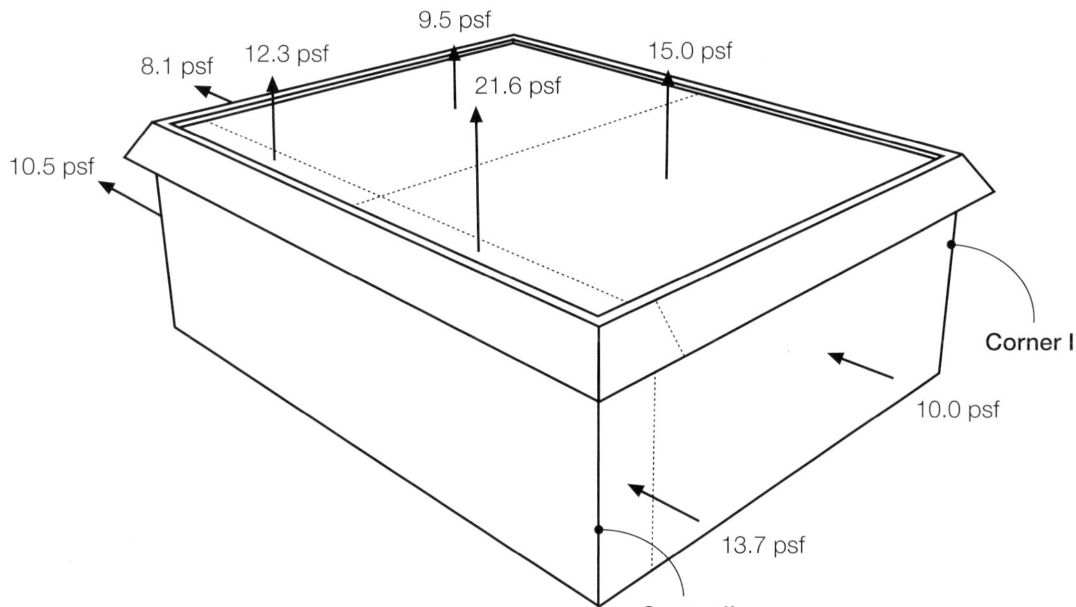

Figure 62b MAIN FRAME PRESSURES–CORNER II (N.E.), CASE B (USING ASCE 7-95 PROCEDURES)

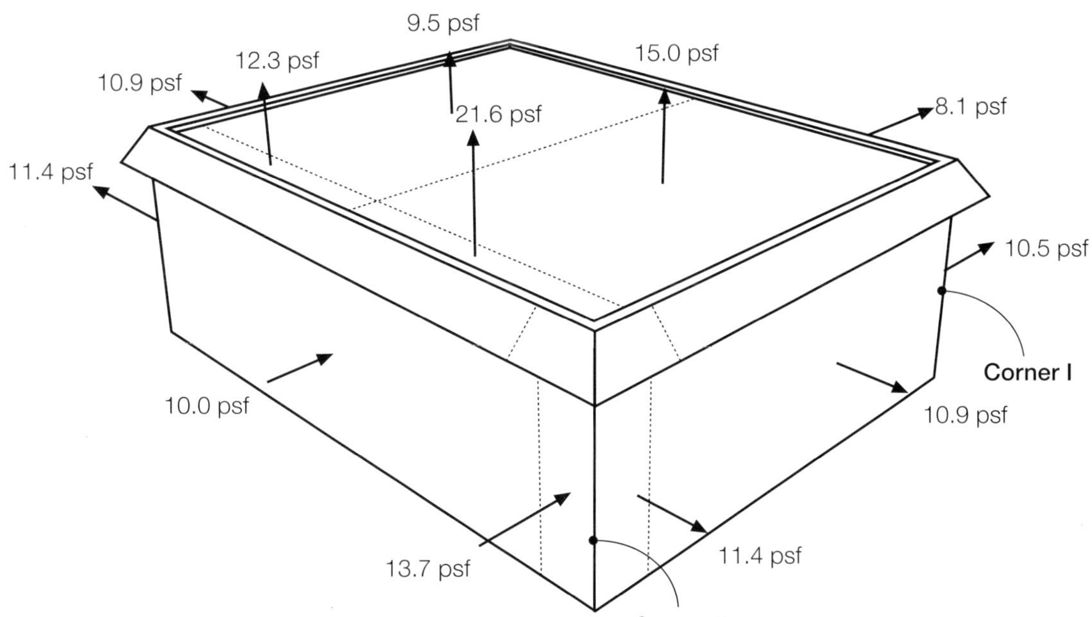

components and cladding, including all pertinent subscripts, is:

$p = q_h[(GC_p) - (GC_{pi})]$

Note that pressure coefficients (GC_p) and (GC_{pi}) are added algebraically before multiplying by the velocity pressure. Subscripts included in the general equation require some additional comment. The velocity pressure q_h is determined at mean roof height for Exposure Category C, regardless of the actual exposure category for buildings of total heights less than or equal to 60 feet (see Section 6.5.3.3 of ASCE 7-95). A more complicated form of the equation is used for buildings of greater height, and the actual exposure category is used.

Velocity Pressure

The velocity pressure q_h, based upon Exposure Category C, is the same for all uses of this equation for the example building and for any building of a height equal to or less than 60 feet. The value is computed from the same general equation for q used in Part 1 of the analysis–specifically:

$q_h = 0.00256\ K_h\ K_{zt}\ V^2\ I$

The coefficient K_h is obtained from Table 6-3 in the referenced standard, Exposure Category C at 30-foot height:

$K_z = 0.98$

As stated in Part 1 of this analysis, the coefficient K_{zt} is not considered here. The calculated value for velocity pressure, then, is:

$q_{30\,ft} = 0.00256\ K_{30\,ft}\ V^2\ I$

$= 0.00256\ (0.98)\ (90)^2\ (1.0)$

$= 20.32$ psf

Observe that this is the same value as calculated for the main-frame in Part 1 of this example and also the same as calculated for the example apartment building analysis, using ASCE 7-93 procedures, even though the equation form is slightly different in ASCE 7-93. This would not necessarily be so for a different importance factor or if K_{zt} were included.

Combined Coefficient (GC_p)

Values for the combined coefficient $(GC_p)$are provided in Figures 6-5A (walls) and 6-5B (roofs) of ASCE-95. Coefficients are obtained from nomographs which are based on the area of the localized surface. Note that this localized area is not the entire wall or roof surface, but only that portion of the subdivided surface being considered. Two coefficients are used for each subdivided area–a positive and a negative for each.

Only areas of subdivisions are needed to determine the combined coefficients, often by interpolation, from the appropriate diagrams of Figure 6-5 in ASCE 7-95. As permitted by the referenced standard

Table 8 APARTMENT BUILDING EXAMPLE–LOCALIZED SURFACE AREAS FOR DETERMINATION OF COMBINED COEFFICIENT (GC_p) FOR COMPONENTS AND CLADDING, ASCE 7-95 PROCEDURES

Building Surface	**Surface Area** (sq.ft.)	(GC_p) (Positive Pressure Coefficient)	(GC_p) (Negative Pressure Coefficient)
1 Roof Inner Area	(80 - 12) (60 - 12) = 2040	+0.2	-0.9
2 Roof Edges	(80 - 12) (6) = 408	+0.2	-1.1
	(60 - 12) (6) = 288	+0.2	-1.1
3 Roof Corners	(6) (6) = 36	+0.25	-2.0
4 Side Walls	(80 - 12) (30) = 2040	+0.7	-0.8
4 End Walls	(60 - 12) (30) = 1440	+0.7	-0.8
5 Side and End Walls	(6) (30) = 180	+0.8	-0.9

(see footnote 5 of Figure 6-5), combined coefficients for walls will be reduced by 10% for values obtained in the nomographs. Pertinent areas for the example building and values for GC_p from Figure 6-5 are shown in Table 8.

Combined Coefficient (GC_{pi})
Values for the combined coefficient (GC_{pi}) are obtained from Table 6-4 of ASCE 7-95 and for this example apartment building and are the same values used earlier for the main-frame pressure equation: +0.18 and -0.18, respectively.

Design Pressures for Components and Cladding
A table format again is used to obtain pressures that apply to components and cladding for this example. All positive and negative pressure combinations of the $[(GC_p) - (GC_{pi})]$ term are shown for each subdivided surface, but only controlling pressures are calculated.

Table 9 provides only the values for coefficients and the calculated pressures obtained from the following equations.

For roof surfaces $p = 20.32\ [(GC_p) \pm 0.18]\ [0.85]$

For wall surfaces $p = 20.32\ [0.9(GC_p) \pm 0.18]\ [0.85]$

Design pressures for components and cladding are summarized in Figure 63. Double arrows are used to indicate that both positive and negative pressures need consideration.

Example 3—Industrial/Commercial Building Wind Pressure Analysis Using ASCE 7-93 Procedures

This example uses a one-story building, gabled rigid-framed in north/south direction and steel braced-frame in east/west direction, as shown in Figure 37. A plan view is not provided since the interior space is free span and not needed for this analysis.

Table 9 APARTMENT BUILDING EXAMPLE–COEFFICIENTS AND CALCULATED DESIGN PRESSURES FOR COMPONENTS AND CLADDING, BASED UPON ASCE 7-95 PROCEDURES

Building Surface	**(GC_p)** (Pressure Coefficient)	**0.9(GC_p)** (10% reduction for walls)	**(GC_{pi})**	**Pressure** (psf)
1	+0.2	–	-0.18	
Roof Inner Area		–	+0.18	**+6.56**
	-0.9		-0.18	**-18.65**
			+0.18	
2	+0.2	–	-0.18	
Roof Edges		–	+0.18	**+6.56**
	-1.1		-0.18	**-22.11**
			+0.18	
3	+0.25	–	-0.18	
Roof Corners		–	+0.18	**+7.43**
	-2.0		-0.18	**-37.65**
			+0.18	
4	+0.7	+0.63	-0.18	
Wall Inner Areas			+0.18	**+13.99**
	-0.8	0.72	-0.18	
			+0.18	**-15.54**
5	+0.8	+0.72	-0.18	
Wall Corners			+0.18	**+15.54**
	-0.9	-0.81	-0.18	
			+0.18	**-17.10**

Assumptions for analysis:

- Suburban location (Exposure Category B)
- The importance factor I is 1.00 (see Table 5, ASCE 7-93)
- Basic wind speed = 100 mph (value from map based upon fastest-mile wind speed)

Design wind pressures will be determined in this analysis. These pressures include those used for design of the main-frame (Part 1) and pressures used for design of individual components and cladding (Part 2).

ASCE 7-93, *Minimum Design Loads for Buildings and Other Structures*, is the source both for procedures and data used in this example. It will be needed for those readers who wish to verify specific procedures and numerical values in the example. However, for convenience, table and figure numbers from the reference are provided as needed again for verification of values and computations.

This example building has a mean roof height less than 60 feet, so procedures for that classification apply to the analysis of its components and cladding. This distinction for building height is not made for the main-frame analysis, and procedures are the same for all buildings regardless of their height.

Part 1–Pressures for Design of the Main-Frame

The velocity pressure q is calculated from the following equation:

$$q = 0.00256\ K_z\ (IV)^2$$

Wind Direction East-West

This wind direction is toward a gabled end of the building and acts parallel with the ridge. Velocity pressures at stepped heights and corresponding values of K_z (from Table 6, ASCE 7-93) are shown to the roof ridge (33 feet) for the windward surface when the wind direction is west to east. Values are the same for wind direction east-to-west, but windward and leeward surfaces are reversed. Note that some pressures on these gabled end walls occur on triangle-shaped and

Figure 63 COMPONENT AND CLADDING DESIGN PRESSURES (USING ASCE 7-95 PROCEDURES)

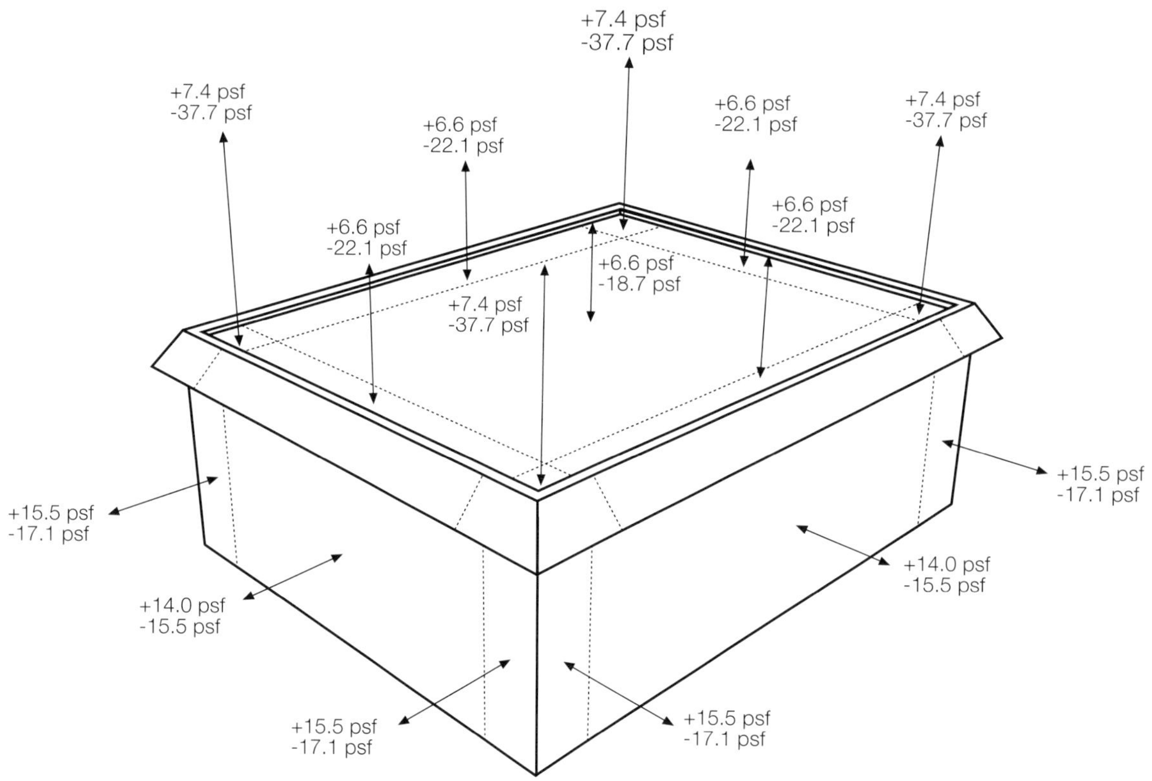

trapezoid-shaped surfaces at heights above the longitudinal side walls.

q_{0-15ft} = 0.00256 (0.37) $[(1.0)(100)]^2$ = 9.47 psf
$q_{15-20ft}$ = 0.00256 (0.42) $[(1.0)(100)]^2$ = 10.75 psf
$q_{20-25ft}$ = 0.00256 (0.46) $[(1.0)(100)]^2$ = 11.78 psf
$q_{25-30ft}$ = 0.00256 (0.50) $[(1.0)(100)]^2$ = 12.80 psf
$q_{30-33ft}$ = 0.00256 (0.52) $[(1.0)(100)]^2$ = 13.31 psf
(K_z is interpolated value)

Pressure coefficients C_p (from Figure 2, ASCE 7-93) for all exterior surfaces depend upon the wind direction, building proportions and–when the wind direction is perpendicular to the ridge–upon roof slope. For the example building, the roof ridge is perpendicular to a north-south wind direction.

Building proportions–Depth of face divided by width of face = L/B = 96/62 = 1.55 > 1. h divided by wall length parallel to ridge = 24/96 < 2.5.

Windward surface:	C_p = + 0.8
Leeward surface:	C_p = - 0.4
Side surfaces:	C_p = - 0.7
Roof surfaces (parallel to ridge):	C_p = - 0.7

Pressures on surfaces are calculated from the following equation.

$p = qG_hC_p - q_h(GC_{pi})$

For the windward surface, the velocity pressure q in the first term on the right side of the equation becomes q_z and values correspond to increments of the stepped pressure profile. For all other surfaces, q becomes q_h, and its value (where h is the average roof height, 15 + 9 = 24 feet) is taken as the same value for $q_{20-25 ft}$ = 11.78 psf (calculated above). The values for G_h are from Table 8, ASCE 7-93, for Exposure Category B. For the windward surface (east-west wind direction) values of z correspond to heights at stepped increments, but for the leeward, side and roof surfaces h is the mean roof height. The values for GC_{pi} (from Table 9, ASCE 7-93) are +0.25 and -0.25 (only that value with an algebraic sign which will produce the largest pressure is used in the following calculations).

<u>Windward wall surface</u>

p_{0-15ft} = 9.47(1.65)(+ 0.8) - (11.78)(-0.25)
= + 15.45 psf
$p_{15-20ft}$ = 10.75(1.59)(+ 0.8) - (11.78)(-0.25)
= + 16.62 psf
$p_{20-25ft}$ = 11.78(1.54)(+ 0.8) - (11.78)(-0.25)
= + 17.46 psf
$p_{25-30ft}$ = 12.80(1.51)(+ 0.8) - (11.78)(-0.25)
= + 18.41 psf
$p_{30-33ft}$ = 13.31(1.49)(+ 0.8) - (11.78)(-0.25)
= + 18.81 psf

<u>Leeward wall surface</u>

p = 11.78(1.55)(- 0.4) - (11.78)(+0.25)
= - 10.25 psf

<u>Side wall surfaces</u>

p = 11.78(1.55)(- 0.7) - (11.78)(+0.25)
= - 15.73 psf

<u>Roof surfaces (both sides, wind parallel with ridge)</u>

p = 11.78(1.55)(- 0.7) - (11.78)(+0.25)
= - 15.73 psf

<u>Wind Direction North-South</u>

Velocity pressure for wind acting perpendicular to the ridge (the long face of the building) is needed for the windward surface only up to the eaves (a height of 15 feet), and does not necessitate a stepped-pressure profile. The corresponding value of K_z (from Table 6, ASCE 7-93) also has the same limit. Since this example building is symmetrically shaped, pressures are the same for wind directions north-to-south and south-to-north, but windward and leeward surfaces are reversed. The same velocity pressure as calculated for east-west wind applies– q_{0-15ft} = 9.47 psf. Velocity pressures for leeward and side walls and roof are based upon h of 24 feet (the mean roof height), and q_h is 11.78 psf (as calculated above for $q_{20-25ft}$).

The gust factor (from Table 8, ASCE 7-95) is based on Exposure Category B, with h = 24 ft is G_h = 1.55.

Pressure coefficients C_p for all exterior wall surfaces depend on the building's proportions in plan. The proportion to be used is the building dimension parallel with wind direction divided by building dimension perpendicular to wind direction, or for the example building with north-south wind direction 62/96 = 0.65. Wall pressure coefficients (from Figure 2, ASCE 7-95) are:

Windward wall surface:	C_p = + 0.8
Leeward wall surface:	C_p = - 0.5
Side wall surfaces:	C_p = - 0.7

Pressure coefficients C_p for all exterior roof surfaces depend upon the building proportion h/L = 24/62 = 0.39 (L is the plan dimension parallel with the wind direction) and upon roof slope (wind direction is perpendicular to roof ridge). For this example building, the roof slope is 30 degrees. Observe in Figure 2 of ASCE 7-93 that the value for the windward roof surface must be interpolated, because tabled values which bracket the *h/L* ratio of 0.39 change from a positive coefficient to a negative coefficient.

Windward roof surface:

C_p = + 0.3 - (0.09/0.2)(0.5) = + 0.08

Leeward roof surface: C_p = - 0.7

The values for *GC* (from Table 9, ASCE 7-93) are +0.25 and -0.25 (only that value with algebraic sign which will produce the largest pressure is used in order to simplify the pressure calculations).

Pressures on surfaces are calculated from the following equation.

$p = qG_hC_p - q_h(GC_{pi})$

Windward wall surface:

p = 9.47(1.55)(+0.8) - 11.78(-0.25) = + 14.69 psf

Leeward wall surface:

p = 11.78(1.55)(-0.5) - 11.78(+0.25) = - 12.07 psf

Side wall surfaces :

p = 11.78(1.55)(-0.7) - 11.78(+0.25) = - 15.73 psf

Windward roof surface:

p = 11.78(1.55)(+0.08) - 11.78(-0.25) = + 4.41 psf

p = 11.78(1.55)(+0.08) - 11.78(+0.25) = - 1.49 psf

(Note that pressure reversal occurs on this surface, so both conditions must be considered.)

Leeward roof surface:

p = 11.78(1.55)(-0.7) - 11.78(+0.25) = - 15.73 psf

Design Pressures: The controlling pressures for design of the main-frame are shown in Figure 64a, a transverse section for north-south wind direction, and Figure 64b, a longitudinal section for east-west wind direction.

<u>Part 2–Pressures for Design of Components and Cladding</u>

Wind direction is not a consideration for this part of the analysis. As was explained in the apartment building example using ASCE 7-93 procedures, all areas of all exterior surfaces are considered when determining pressures on components and cladding. Wall and roof surfaces are subdivided into smaller areas for which different combined pressure/gusting coefficients apply. There are two subdivisions that apply to wall surfaces: corners and inner areas. There are three subdivisions that apply to roof surfaces: corners, edges and inside areas. All other commentary in the apartment building example analysis based upon ASCE 7-93 also apply to this example. To shorten the work, only those calculations necessary to obtain design pressures are included.

Since *h* = 24 feet, which is less than 60 feet, Figure 3 of ASCE 7-93 is used to obtain the combined coefficient GC_p. Exposure Category C is to be used regardless of actual exposure category, so K_h = 0.92.

Figure 64a MAIN-FRAME PRESSURES AT LONGITUDINAL SECTION (USING ASCE 7-93 PROCEDURES

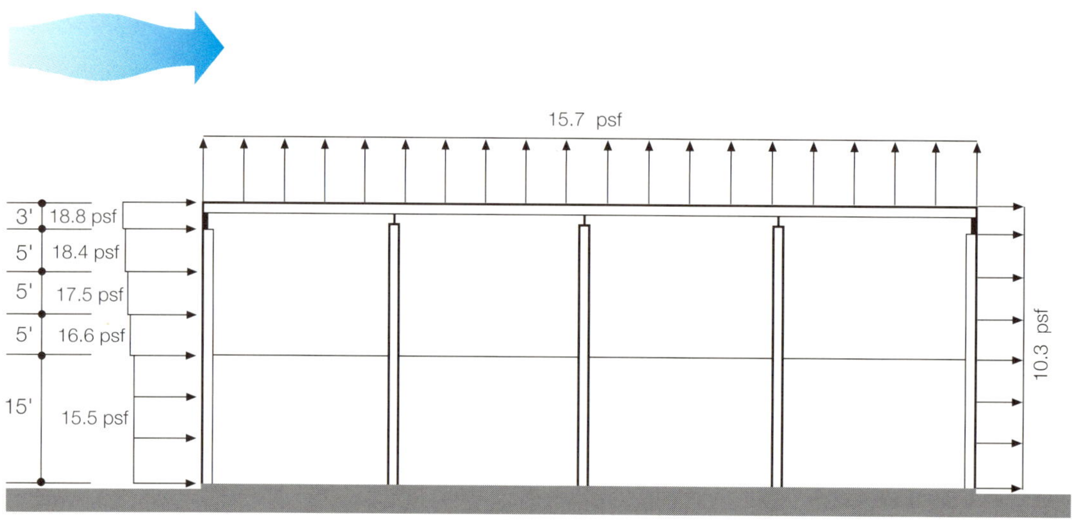

q_h = 0.00256 (0.92) [(1.0)(100)]^2 = 23.55 psf

GC_{pi} = + 0.25 and - 0.25 (same as for main-frame analysis)

Width of localized area strip, a:

0.10(62) = 6.2 ft

0.4(24) = 9.6 ft

Use a = 6.2 ft

Pressures on localized areas are determined from the following general equation (ASCE 7-93, Table 4):

$p = q_h[(GC_p) - (GC_{pi})]$

Pressure coefficients (GC_p) and (GC_{pi}) are added algebraically before multiplying by velocity pressure.

Pressures are as follows (see Figure 27 for surface designations):

<u>Side walls (L = 96 ft)</u>

Surface 4 — A = 15(83.6) = 1254 sq ft

GC_p = +1.0 and -1.1

p = 23.55(+1.0) - 23.55(-0.25) = + 29.44 psf

p = 23.55(-1.1) - 23.55(+0.25) = - 31.80 psf

Surface 5 — A = 6.2(15) = 93 sq ft

GC_p = +1.2 and -1.5

p = 23.55(+1.2) - 23.55(-0.25) = + 34.15 psf

p = 23.55(-1.5) - 23.55(+0.25) = - 41.22 psf

<u>End walls (L = 62 ft)</u>

Surface 4 — A = 49.6 [(33 + 18.6)/2] = 1280 sq ft

GC_p = +1.0 and -1.1

p = 23.55(+1.0) - 23.55(-0.25) = + 29.44 psf

p = 23.55(-1.1) - 23.55(+0.25) = - 31.80 psf

Surface 5 — A = 6.2 [(15 + 18.6)/2] = 104 sq ft

GC_p = +1.2 and -1.5

p = 23.55(+1.2) - 23.55(-0.25) = + 34.15 psf

p = 23.55(-1.5) - 23.55(+0.25) = - 41.21 psf

<u>Roof (L = 96 ft, width = 36.7 ft, Θ = 30°)</u>

Surface 1 — A = (96 - 12.4)(36.7 -12.4) = 2032 sq ft

GC_p = -1.1

p = 23.55(-1.1) - 23.55(+0.25) = - 31.80 psf

Surface 2 — A = 6.2 (96 - 12.4) = 518 sq ft

GC_p = -2.0

p = 23.55(-2.0) - 23.55(+0.25) = - 52.99 psf

Surface 3 — A = 6.2(6.2) = 38 sq ft

GC_p = -2.5

p = 23.55(-2.5) - 23.55(+0.25) = - 64.76 psf

A summary of these localized pressures is illustrated in Figure 65.

Example 4—Industrial/Commercial Building Wind Pressure Analysis Using ASCE 7-95 Procedures

This example continues to use a one-story building with a gabled rigid-frame in north/south direction and steel braced-frame in east/west direction as shown in Figure 35 (the same building for which ASCE 7-93 procedures were used for analysis).

Assumptions for analysis:

Figure 64b MAIN-FRAME PRESSURES AT TRANSVERSE SECTION (USING ASCE 7-93 PROCEDURES)

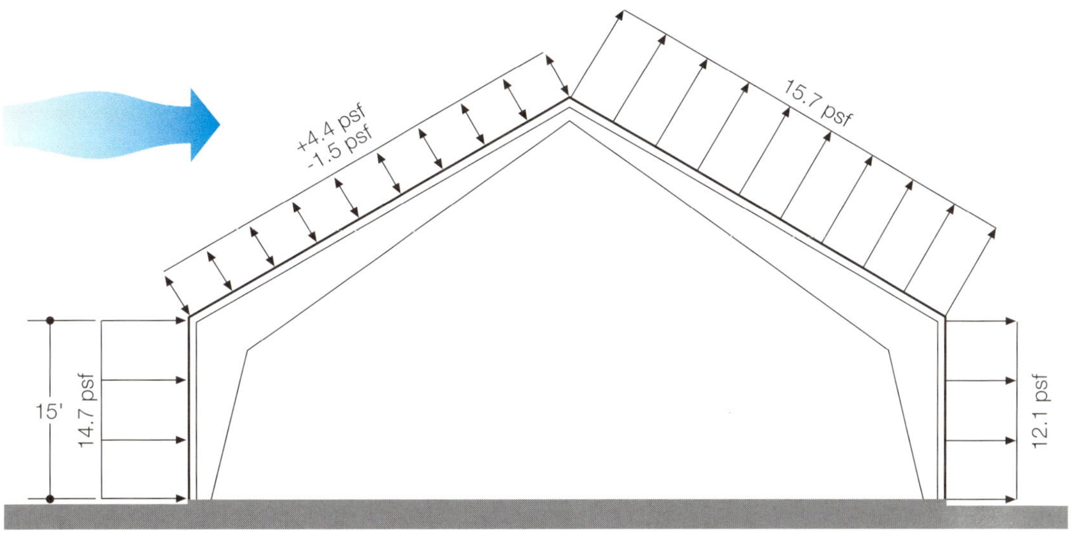

- Suburban location (Exposure Category B)
- Importance factor I = 1.0
- Basic wind speed = 130 mph (value from map based upon three-second gust wind speed)

The building for this example analysis is in the same location as the building of the example analysis using ASCE 7-93 procedures. While ASCE 7-93 procedures for a coastal hurricane region, have a fastest-mile wind speed velocity of 100 mph, the same location under ASCE 7-95 procedures has a basic wind speed of 130 mph (three-second gust wind speed).

Design wind pressures for the example building are determined in this analysis. These include pressures for design of the main-frame (Part 1) and pressures which are used for design of individual components and cladding (Part 2).

ASCE 7-95, *Minimum Design Loads for Buildings and Other Structures*, is the source both for procedures and data used in this example. Consult this reference if you wish to verify specific procedures and numerical values used in this example. For your convenience, table and figure numbers from ASCE 7-95 are provided as needed in this example.

Wind pressure analysis for this example building, which has a height of 33 feet, is based on the ASCE 7-95 prescribed procedure for buildings up to 60 feet high. The example building, with a mean roof height h = 24 feet, has an aspect (height to least width) ratio of h/w = 24/62 = 0.387 (which is less than 1.0) and meets the qualifications for analysis as a low-rise building.

Calculated Velocity Pressure: The equation for velocity pressure q in this example is:

$q_h = 0.00256\, K_h\, K_{zt}\, V^2\, I$

Following the procedure for "low-rise buildings," Exposure Category C is used regardless of actual exposure category. For h = 24 ft, K_h = 0.93 (Table 6-3, ASCE 7-95), but calculated pressures may be multiplied by 0.85 when the building is sited in an Exposure Category B location. Since no local terrain data were given, assume K_{zt} = 1.0. The calculated value for q is:

$q_h = 0.00256\,(0.93)\,(1.0)\,(130)^2\,(1.0) = 40.24$ psf

Figure 65 COMPONENT AND CLADDING DESIGN PRESSURES (USING ASCE 7-93 PROCEDURES)

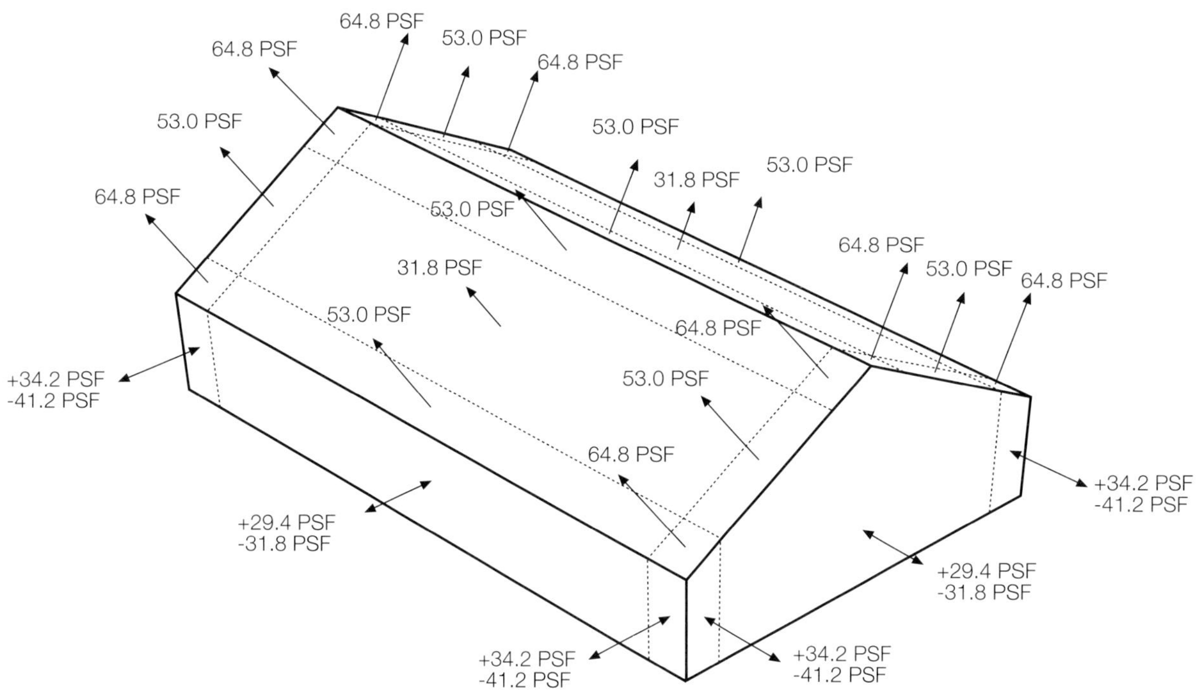

Part 1. Pressures for Design of the Main-Frame

The pressure equation for "low-rise buildings" is:

$p = q_h[(GC_{pf}) - (GC_{pi})]$

Due to the example building's symmetry, complete pressure analysis for the main-frame can be accomplished by considering just two adjacent (not diagonally opposite) corners of the building, with two cases (Case A and Case B) considered for each corner. Controlling design pressures can be derived from these conditions without analysis of the other two corners.

The values for (GC_{pi}) from Table 6-4 that apply to the example building are:

(GC_{pi}) = +0.18 and -0.18

In accordance with footnote 4 of Table 6-4, design pressures may be multiplied by 0.85 for buildings located in Exposure Category B.

Corner I (see Figure 66 for location of this corner)

For the example building, Case A values for (GC_{pf}) obtained from Figure 6-4, ASCE 7-95 (roof slope θ = 30°), are based upon the surfaces as labeled in Figure 66 for Corner I, Case A (wind direction perpendicular to ridge).

Different values of (GC_{pf}) are used for Corner II, Case A (wind direction parallel with ridge). It is essential to correlate references to building surfaces, as required for obtaining values for (GC_{pf}), with actual building surfaces as each corner is considered. The dimension a which defines the extent of surfaces is based on the same criteria as in ASCE 7-93 (a = 6.2 ft, and $2a$ = 12.4 ft.

Calculated Pressures–Corner I

Surface pressures are calculated using the following equation and the appropriate combined pressure coefficient (GC_{pf}) value from those in Figure 6-4, ASCE 7-95. Of the two possible combinations of terms involving (GC_{pi}) = ±0.18 (Table 6-4, ASCE 7-95), only the controlling pressure is calculated. Calculations reveal that no surface has both a positive and a negative pressure, so the larger pressure magnitude is used.

$p = 40.24\,[(GC_{pf}) \pm 0.18][0.85]$

Case A:

Surface 1	(GC_{pf}) = + 0.56	p = + 25.31 psf
Surface 2	(GC_{pf}) = + 0.21	p = + 13.34 psf
Surface 3	(GC_{pf}) = - 0.43	p = - 20.86 psf
Surface 4	(GC_{pf}) = - 0.37	p = - 18.81 psf
Surface 1E	(GC_{pf}) = + 0.69	p = + 29.76 psf
Surface 2E	(GC_{pf}) = + 0.27	p = + 15.39 psf
Surface 3E	(GC_{pf}) = - 0.53	p = - 24.28 psf
Surface 4E	(GC_{pf}) = - 0.48	p = - 22.57 psf

Case B:

Surface 1	(GC_{pf}) = - 0.48	p = - 21.55 psf
Surface 2	(GC_{pf}) = - 0.69	p = - 29.76 psf
Surface 3	(GC_{pf}) = - 0.37	p = - 18.81 psf
Surface 4	(GC_{pf}) = - 0.45	p = - 21.55 psf
Surface 5	(GC_{pf}) = + 0.40	p = + 19.84 psf
Surface 6	(GC_{pf}) = - 0.29	p = - 16.08 psf
Surface 1E	(GC_{pf}) = - 0.48	p = - 22.57 psf
Surface 2E	(GC_{pf}) = - 1.07	p = - 42.76 psf
Surface 3E	(GC_{pf}) = - 0.53	p = - 24.28 psf
Surface 4E	(GC_{pf}) = - 0.48	p = - 22.57 psf
Surface 5E	(GC_{pf}) = + 0.61	p = + 27.02 psf
Surface 6E	(GC_{pf}) = - 0.43	p = - 20.86 psf

A summary of design pressures for Corner I is provided in Figure 66.

Corner II (See Figure 67 for location of Corner II)

Case A (roof slope θ = 0 degrees, see footnote 2, Figure 6-4, ASCE 7-95): Values for the combined coefficient (GC_{pf}) for this case, where the wind direction is parallel with the roof ridge, are based on a roof slope θ = 0. Hence, the coefficients are different than those used for Corner I and are taken from a different row of the same table.

Case B: Values for (GC_{pf}) are the same as for Corner I, Case B, except that surfaces must be correlated with the rotated building ; e.g., Surface 1 for Corner II, Case B is Surface 5 for Corner I, Case B.

Calculated Pressures–Corner II

$p = 40.24\,[(GC_{pf}) \pm 0.18][0.85]$

Case A:

Surface 1	(GC_{pf}) = + 0.40	p = + 19.84 psf
Surface 2	(GC_{pf}) = - 0.69	p = - 29.76 psf
Surface 3	(GC_{pf}) = - 0.37	p = - 18.81 psf
Surface 4	(GC_{pf}) = - 0.29	p = - 16.08 psf
Surface 1E	(GC_{pf}) = + 0.61	p = + 27.02 psf
Surface 2E	(GC_{pf}) = - 1.07	p = - 42.76 psf
Surface 3E	(GC_{pf}) = - 0.53	p = - 24.28 psf
Surface 4E	(GC_{pf}) = - 0.43	p = - 20.86 psf

Case B: Values are the same as for Corner I, Case B, when surfaces are correlated.

Figure 66a MAIN-FRAME PRESSURES–CORNER I (S.W.), CASE A (USING ASCE 7-95 PROCEDURES)

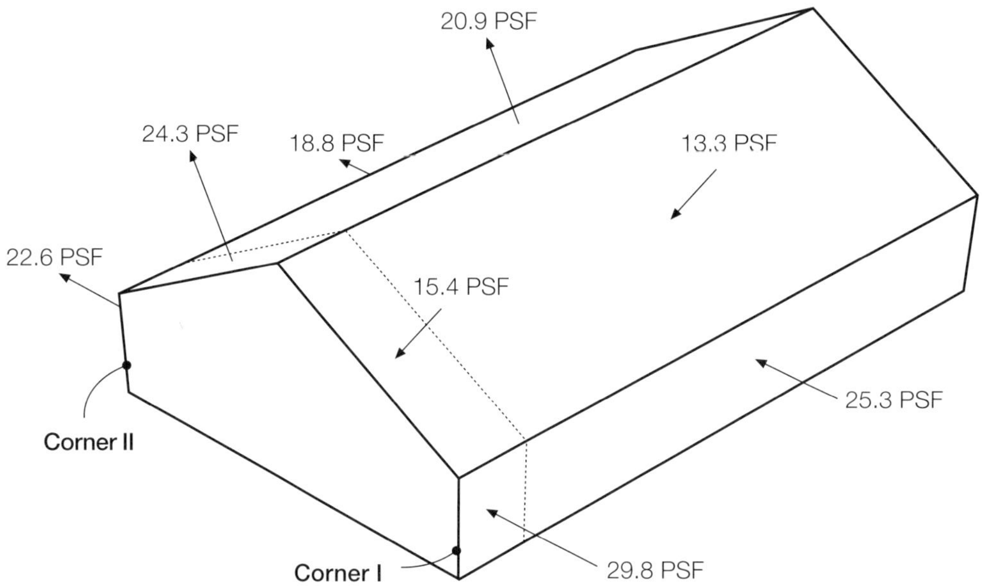

Figure 66b MAIN-FRAME PRESSURES–CORNER I (S.W.), CASE B (USING ASCE 7-95 PROCEDURES)

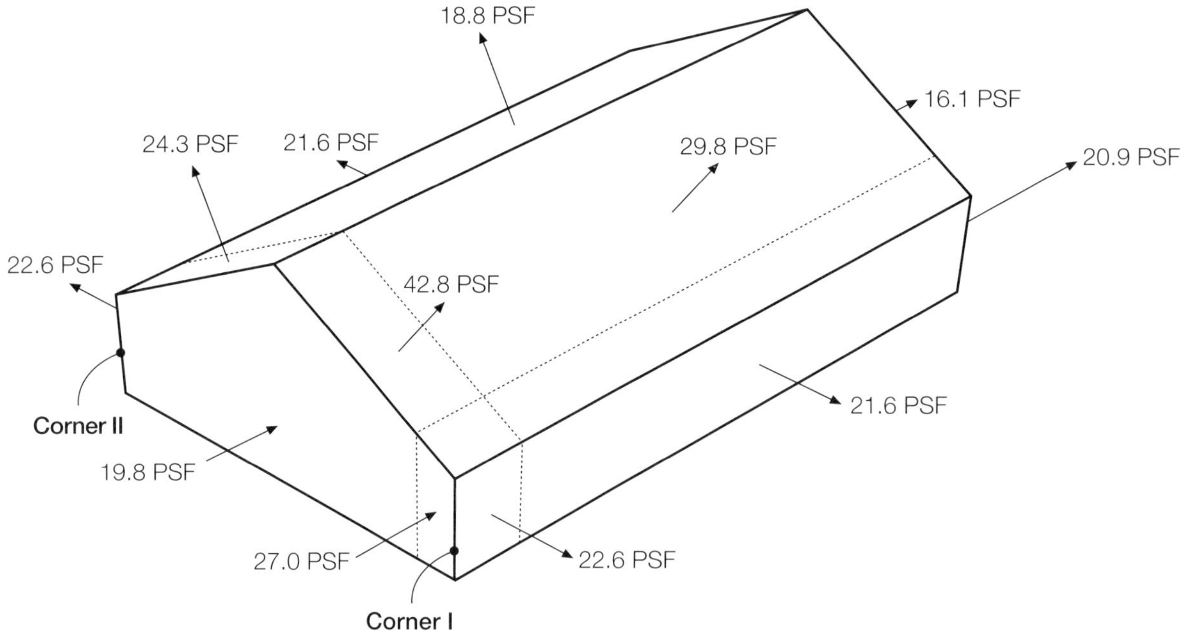

Figure 67a MAIN-FRAME PRESSURES–CORNER II (S.E.), CASE A (USING ASCE 7-95 PROCEDURES)

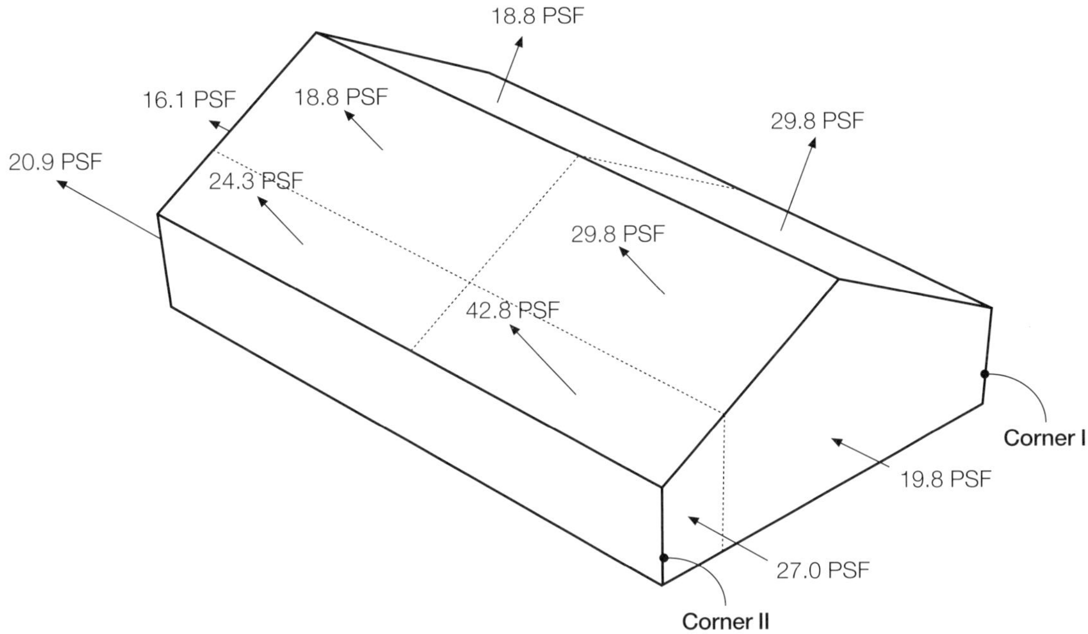

Figure 67b MAIN-FRAME PRESSURES–CORNER II (S.E.), CASE B (USING ASCE 7-95 PROCEDURES)

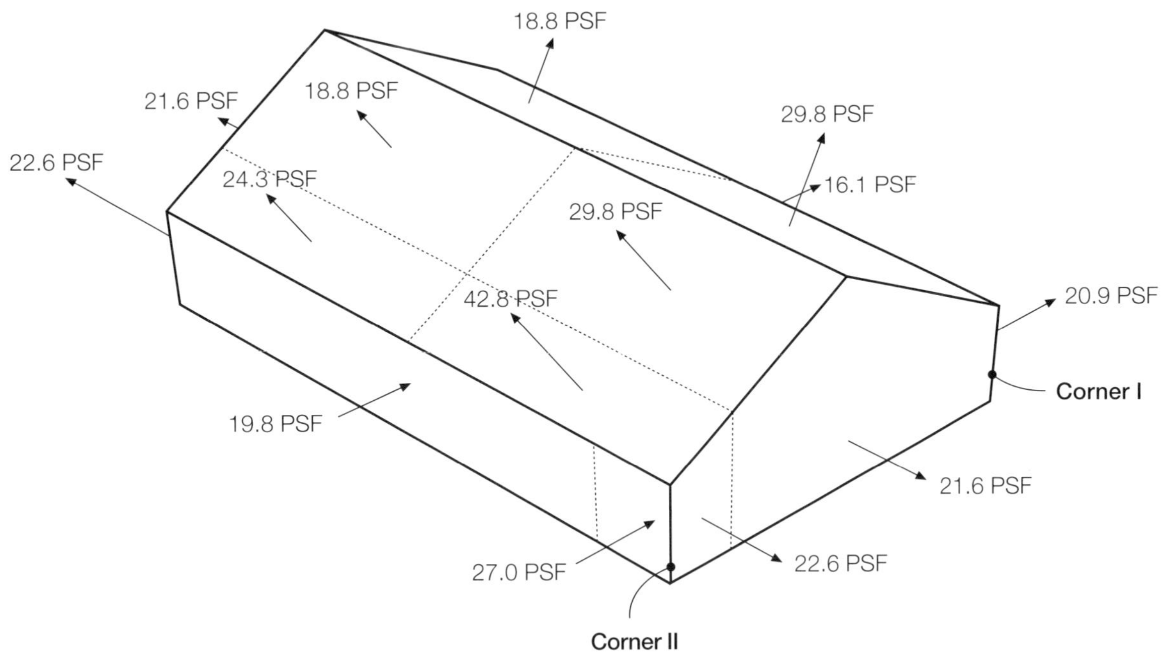

Design Pressures

Controlling pressures for design of the main-frame are shown in Figure 66 (Corner I, Cases A and B) and Figure 67 (Corner II, Cases A and B). All four pressure loadings must be considered in design of the main-frame. Overall building torsion as a consequence of pressure distribution must be considered in the loadings. The main-frame must be designed for the most critical condition, which would require four separate analyses. If computer programs are not readily available for this task, most designers will undertake an overall conservative simplification procedure, using larger forces than produced by the design pressures.

Part 2. Pressures for Design of Components and Cladding

Localized surfaces for which increased pressures are considered for the example building are prescribed in Figure 6-5 of ASCE 7-95 ($h < 60$ ft).

The general form of the pressure equation for components and cladding is:

$p = q_h[(GC_p) - (GC_{pi})]$

q_h = 40.24 psf (same value as in Part I).

(GC_p) is to be determined for each localized surface (Figure 6-5, ASCE 7-95).

$(GC_{pi}) = \pm 0.18$

a = 6.2 ft

Both terms within the brackets of the general equation may be multiplied by 0.85. The resulting equation from which pressures are calculated is:

For roof surfaces — $p = 40.24\ [(GC_p) \pm 0.18]\ [0.85]$

For wall surfaces — $p = 40.24\ [(GC_p) \pm 0.18]\ [0.85]$

Based upon these equations, controlling pressures for localized surfaces are determined in Table 10 and pressures for components and cladding are summarized in Figure 68.

STRENGTHENING BUILDINGS FOR WIND RESISTANCE

Construction details and practices affecting building wind resistance are examined in this section. Concepts

Table 10 INDUSTRIAL/COMMERCIAL BUILDING EXAMPLE–COEFFICIENTS AND CALCULATED DESIGN PRESSURES FOR COMPONENTS AND CLADDING, BASED UPON ASCE 7-95 PROCEDURES

Building Surface	Surface Area (sq.ft.)	(GC_p) (Pressure Coefficient)	(GC_{pi})	Pressure (psf)
1	(96 - 12.4) (35.8 - 12.4) = 1956	+0.3	-0.18	+16.42
Roof Inner Area		-0.8	+0.18	-33.52
2	At Side Walls	+0.3	-0.18	+16.42
Roof Edges	(96 - 12.4) (6.2) = 518	-1.4	+0.18	-54.04
	At End Walls	+0.3	-0.18	+16.42
	(35.8 - 12.4) (6.2) = 145	-1.4	+0.18	-54.04
3	(6.2) (6.2) = 38	+0.4	-0.18	+19.84
Roof Corners		-1.7	+0.18	-64.30
4	Side Walls	+0.7	-0.18	+30.10
Wall Inner	(96 - 12.4) (15) = 1254	-0.8	+0.18	-33.52
Areas	Gabled End Walls	+0.7	-0.18	+30.10
	(62 -12.4) (33 + 18.6)/2 =1277	-0.8	+0.18	-33.52
5	Side Walls	+0.8	-0.18	+33.52
Wall	(6.2) (15) = 93	-1.1	+0.18	-43.78
Corners	Gabled End Walls	+0.8	-0.18	+33.52
	(6.2) (15 + 18.6)/2 = 104	-1.1	+0.18	-43.78

examined are relevant for both new and existing construction. Applications may be different, however, because of inherent problems either in accessing or altering components in construction already completed. Improvements for wind resistance in existing buildings often create tougher design and engineering challenges than for new projects. Achieving acceptable wind resistance in new buildings is actually a relatively simple task, especially because needed structural elements, strength requirements, and connection details can be incorporated during the design process. For existing buildings that are found to have deficiencies for wind resistance, there often is no easy correction of the deficiencies without damaging finishes. This is particularly so for those deficiencies of the main structural system intended to resist the wind forces.

Rarely are any two buildings identical, even those which employ the same primary structural concepts. Clearly, there are many ways to develop shear-wall structural action, many ways to frame a structure and thus affect the load path, and many ways to provide suitable diaphragms. As well, there are many ways to provide for proper force development in connections of assemblies, for example, tension, bearing or shear connections are all possible with nails, bolts, welding or other fasteners. Due to all of these possible variations, there are only general rules we might suggest which address all conditions involving wind resistance. Beyond these general rules, specific conditions must be addressed.

Evaluation–A Necessary Beginning Point

There is one activity for assessment of wind vulnerability that applies to all buildings, which should now be obvious. Evaluation of wind resistance for every building requires a wind analysis uniquely for that building.

Analysis of a building being designed is based on construction systems, details and dimensions tentatively established but usually changeable. Deficiencies that might be discovered through analysis can be corrected before construction.

However, analysis and its application are necessarily

Figure 68 COMPONENT AND CLADDING DESIGN PRESSURES (USING ASCE 7-95 PROCEDURES)

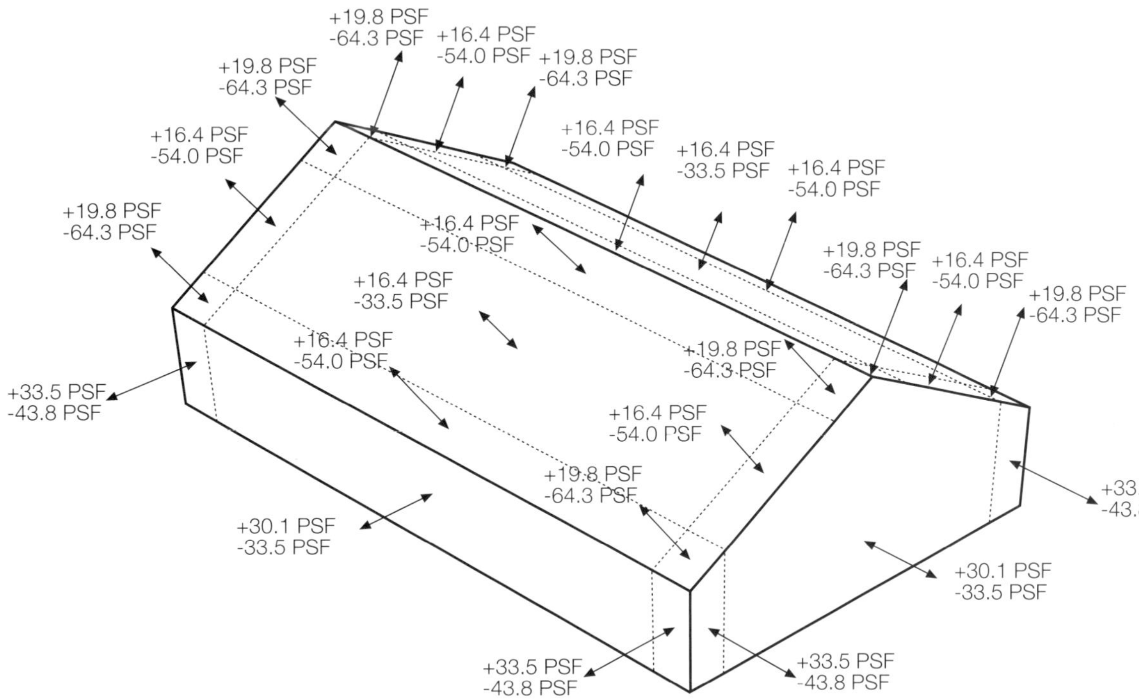

based upon assumptions–of configuration, strengths of components, assembly, and connections. The constructed building must satisfy these assumptions if the analysis and results are to remain valid. Design coordination among architects, analysts, those who prepare the construction details and on-site inspections is necessary to ensure that design assumptions and end product match.

Analysis of an existing building must include field observations, either close inspection of pertinent portions of the building or, with the good fortune of having construction drawings available, careful review of those documents. There is no other way that deficiencies can be identified.

Complete reliance upon construction documents may not be wise, however. Actual connections and fasteners may not match those shown in construction drawing details. Even when a particular building's wind-resistant capacity is examined, important considerations may be hidden within assemblies or behind finishes. It has been stated in earlier parts of this monograph that connection details are a significant consideration for proper wind resistance. If these are hidden from view, there is no way to know their strength unless exposed to visual examination.

Priorities for Strengthening

In order to provide helpful guidance for improving building resistance to wind, yet recognizing that construction variations are too extensive to allow comprehensive treatment, we have chosen to highlight important issues by drawing upon experiences from wind-damaged buildings. Actual wind damage, both in type and observed frequency establish the priorities for this discussion.

Consider buildings damaged in Hurricanes Andrew and Iniki, which were mentioned earlier. Extensive field observations were documented for both hurricanes owing to widespread building damage of all degrees, ranging from minor tearing to devastation.

As a point of departure, several overall observations are summarized regarding wind damage in the impacted regions:

- Damage was most extensive, in the number of affected buildings and the degree of damage, to wood-frame and wood-frame with masonry buildings, comprising mostly residential buildings and multi-unit buildings.
- Damage, to whatever degree, was of the progressive failure type, meaning that failure of lesser important components led to weakening of the more important building elements, with structural collapse an end result in some cases.
- Property losses were made worse by water damage occurring when a building's enclosure envelope was lost.
- Damage to commercial buildings was largely related to failures of glazing, although cladding failures were experienced as well, some of which led to complete structural collapse.
- Damage to large-span, light metal-framed industrial buildings was disproportional to their number.

Building damage resulting from these hurricane winds was due to a number of causes; it occurred on many types of building elements and became progressively worse in many buildings for different reasons. These causes, building elements, and reasons for progressively worse damage are separately summarized here and provide clues as to where strengthening techniques can most effectively be applied.

Causes for initial damage are largely attributed to:

- Glazing failure and failures at other types of openings.
- Pull-out of fasteners and tearing of materials from fasteners.
- Missing fasteners.
- Wind-borne debris.

Building elements most frequently affected include:

- Windows and doors, including glazing, frames, and doors.
- Roof coverings, including shingles, tiles, and underlayments.
- Wood sheathing and cladding.
- Wall plates or other elements providing stability at tops of walls.
- Metal decking and metal cladding.
- Masonry walls stabilized by other building elements.

Reasons for consequent damage–involving both development of wind pressures and structural behavior–include:

- Abrupt pressure changes when openings fail.
- Loss of essential elements in construction assemblies.

- Loss of lateral support for elements, including those which are part of a main-frame.

Strengthening Considerations and Methods

With the causes of wind damage and these consequent effects on buildings in mind, the remainder of this section is given to their more extensive examination. Causes, effects, and reasons for failure are treated together for the issues and elements that are discussed. As well, methods are suggested for strengthening these possible weak points in buildings.

Windows and Doors

Progressive damage is often initiated by failures of windows and doors. When glazing material shatters, building interiors are subjected to internal pressures as discussed previously. Failure of openings is usually abrupt and results in rapid pressure loading of building elements. In some instances the change in loading will cause stress reversal; in other cases loading may be increased; in still other cases, particularly building interiors, elements not previously loaded will experience pressure loadings.

Glazing failure is most often associated with missile damage from the extensive amount of wind-blown debris that accompanies winds of high velocity. But broken glazing is not the only type of initial failure. Window and door frames also have been observed to tear loose from adjoining framing. These failures occur primarily when the adjoining framing is wood. Increased use of light-gauge metal framing studs, currently happening because of the high cost of wood, may result in some failures for this condition in the future as well, unless anchorage is given more attention. Indeed, screws pulling out of light-gauge metal framing has been observed. Entry door failures, even when solid, also occur, usually at the door's mid-height on the hinge side. Sliding glass doors are especially vulnerable, although usually through glass failure.

Not much can be done to remedy potential glass failure, short of either replacement with missile resistant glazing (products are becoming available to meet the stringent performance requirements introduced in South Florida), or adding shuttering that can be closed when needed to provide protection.

Secure anchorage of window and door frames can be accomplished with adequate nailing. Too often, minimum nailing is used, most often for alignment rather than strength. Additional nailing at jambs of door frames can be accomplished easily for existing buildings because the locations are typically readily accessible. This is less likely for window frame anchorage in existing buildings–particularly in wood framing. Most prefabricated window frames are only minimally anchored from the outside face, after which finish cladding and trim are added. In such situations, trim may need to be removed before additional fasteners are placed.

Roof Membrane and Sheathing

Progressive failure following the initial failure of roofing membranes and substrate led to major structural damage of buildings in Hurricanes Andrew and Iniki. Reasons for this are attributed to construction practices later found to be especially vulnerable. In particular, roof trusses, especially those at the ends of gabled roofs, commonly have been framed in a manner that makes their lateral stability dependent upon participation of sheathing, or upon an intact gable-end wall, for force resistance. If the roof edge sheathing tears away, the roof truss lateral stability is lost in such cases. If the gable-end wall fails first, the sheathing is effected almost immediately by additional uplift forces. Extensive, if not total, roof collapse follows. Photo 3 on page 2 near the beginning of this monograph shows this kind of complete roof failure.

Little can be done to eliminate damage to shingles and roof coverings commonly installed on light-framed buildings with sloped roofs. Often, composite shingles and roofing felt simply will tear away. Roofing nails with broad heads perform better than staples in resisting tearing, but membrane failures occur for both types. Ordinary wood shingles will also tear loose. Damage reports on the two hurricanes do not address the performance of single-ply membranes. Heavier roofing materials, such as tiles, have also been blown loose in extreme winds. Too often, only gravity is relied on to hold these in place. Even in installations where the tiles are anchored, the common technique is to secure them with one, or at most two nails, into the sheathing. Based upon observed damage, this is simply not enough. A report on damage from Hurricane Andrew indicates that roofing tiles set in mortar beds, with mortar as the only anchorage, was the least successful method of securing roof tiles.

The principal problem associated with initial roof

failure, however, is inadequate sheathing anchorage–especially at roof edges, as discussed earlier. Increased wind pressures at these edges, as described earlier in the monograph, is one reason. A second reason is simply insufficient nailing, a situation easily remedied by improved construction practices and on-site inspections. Photo 6 shows a piece of plywood sheathing found lying on the ground following Hurricane Andrew. If one looks closely, the rafter line is visible, as is a line of staple fasteners. Construction workers missed the rafter, and it can be seen that the plywood lifted off the roof because of this. Such inferior workmanship reinforces the essential need for quality construction and proper inspection.

Reduced possibility of sheathing failure is accomplished by sufficient anchorage to the collector elements. Nailing patterns, as suggested in building standards, must be adhered to. Problems involving sheathing anchorage in existing buildings are not easily identified and corrected until roofing material is replaced. Since periodic replacement of most roofing materials occurs, it is the best time for inspection and connection of existing sheathing anchorage. Nail heads are easy to see, and the nailing pattern in place can be compared with what is recommended.

Another problem of gable-end wall vulnerability involves the roof sheathing. It is common construction practice to use roof sheathing as the means of truss top chord bracing. This sheathing occurs at edges of increased wind pressure where failure is expected to occur first. When the sheathing is blown away, lateral support for end trusses is lost. To ensure that roof trusses maintain their position regardless of any loss of stabilizing elements (for example, the plywood sheathing), roof trusses need bracing in their transverse direction (besides any that may accrue from the sheathing), or steps need to be taken to ensure that the gable-end wall remains in place.

Wall Plates

Several problems associated with wind resistance arise involving top plates for walls. These plates act as eaves

Photo 6

Photo courtesy of the Federal Emergency Management Agency

beams (also discussed in the section dealing with building design concepts to resist wind). They are a part of the main-frame in light-wood construction, as well as light-metal construction, and they are essential for a complete and adequate load path.

The need was discussed earlier in the apartment building example for continuity of wall plates which serve as collectors. Wind loadings typically are lateral, so wall plates acting as collectors receive their loading from or transfer loads to vertically spanning members (such as studs). These are occasions when the plates are loaded horizontally, in addition to any gravity loads that might be carried. A discontinuous plate cannot transfer the horizontal bending effect to the vertical members. Double plates can act in bending provided that their splices do not occur at the same location or in near proximity to each other.

It was also noted in the apartment building example that force transferred from a wall-plate collector to the next element in the load path sequence–often a shear wall–requires a secure connection at the reaction location. For two walls intersecting at a corner when a single wall plate is used, one of the two plates cannot be firmly secured to the corner studs by ordinary nailing techniques. If a double plate is used, plates can be overlapped at the corner and thereby permit a secure connection for each of the two corner walls. Without suitable anchorage at the corner top, wind pressures on the separate surfaces meeting at the corner may cause the corner to separate, at which time the walls no longer brace each other. Wall plates at all exterior walls may be vulnerable.

As is implied in this description of wall plates, there are several possible locations for structural weakness, any one of which may lead to serious structural damage. Unfortunately, wall plates in existing construction are hidden by interior and exterior finishes. It's even difficult to determine whether a single plate or a double plate has been used. Hence any remedial action to identify and correct these possible deficiencies may entail considerable demolition of finishes. Since such deficiencies can lead to major structural

Photo 7

Photo courtesy of the Federal Emergency Management Agency

damage, plate continuity between supports and secure anchorage at supports is a serious issue; but we can suggest no simple "fix". One possible solution is use of partitions placed perpendicular to the wall having a discontinuous plate. Such partitions then become shear walls and require their own strengthening, using techniques such as let-in bracing or suitable sheathing.

Wall plates at gable ends present less evident potential problems. The usual construction practice is to frame walls on all sides to eaves height, and then to infill with studs from plate height to the sloping roof at gable ends. The result of this framing practice is that the eaves-level plates on the gable end are loaded from wind pressures both above and below the plate level. Hence these end-wall plates will carry larger loading per lineal foot than will wall plates for the sides. Photo 7 illustrates the severity and damaging effect of this loading condition.

In a report on building damage and performance following Hurricane Andrew, the American Plywood Association (APA) suggests new detailing practices which deal with the gable-end wall strengthening. Another method for strengthening gable-end walls is suggested in a publication titled *Building Performance: Hurricane Andrew in Florida*, prepared by the Federal Emergency Management Agency. Both methods involve changes in construction practices to ensure bracing of the end wall at or near the level of the wall plate.

It is noteworthy that modular wood-frame construction–that is, the pre-fabrication of modular elements at a factory or building site which then are anchored together for a complete building–has been found to provide better resistance to wind than on-site construction. The main reason is that additional framing members are provided at edges of modular units that later connect to other units. Hence continuity and additional strength at plates are provided simply to facilitate on-site assembly.

Manufactured Connectors

A variety of types of metal devices for wood-framing connections and ties are readily available. Properly used, they are especially effective for strengthening construction against wind effects. Rather than illustrating these connectors, comments are directed to their proper use for wind conditions.

A fundamental principle to be satisfied is that the building behave as a single unit when subjected to lateral forces. One aspect of such desirable behavior is that load paths be complete. This sometimes is referred to as "tying the building together" and entails consideration of uplift forces as well as gravity forces. Manufactured connectors can contribute to achieving this for wood-framed systems. But they must be the right type for each particular connection, so that forces are transferred through connections to the next member in the load-path sequence.

Locations where continuity of load path is often compromised are at roof-to-wall plate connections and floor plate-to-foundation connections. At these locations, several important structural members meet, and forces due to loading are transferred, usually toward the foundations. Framing assemblies at these locations are typically done under gravity-loading conditions, and what seems right for gravity loading may be inadequate for lateral loading. Gravity loads typically produce compression forces which are readily accommodated in most connection details, but uplift usually produces tension forces that ordinary detailing and fasteners do not readily accommodate. Two simple examples are uplift on a roof joist or rafter, and uplift at a sill. Manufactured connectors can provide the needed resistance against the uplift that ordinary nailing cannot.

Continuity of load path through connections requires secure ties from rafters to the wall framing members, and from wall framing members to foundations. Little is accomplished by anchoring rafters to the top wall plate if that plate is not securely tied to wall framing members. Anchor bolts connecting sills to foundations are also of little help if wall framing members are not secured to the sill plate. To provide the needed anchorage at these and other locations, metal strap ties often are used to tie elements together above and below the plates. Strap ties are the most economical way to achieve the needed continuity, although framing anchors connecting each and every member to the next also can be used.

Masonry Walls

Masonry walls can be constructed to develop bending resistance either vertically or horizontally. An example of vertical bending is a wall spanning from a foundation to the floor above (floor to floor) and subjected to wind pressure on the wall surface. A wall cantilevered from a foundation such as a garden wall also needs to develop bending resistance vertically. An example of horizontal bending is a wall, again subjected to wind pressure, with its lateral support provided by other walls positioned perpendicular to the loaded wall, so that horizontal reaction forces are developed at the other walls. Reinforcement requirements are different for vertical and horizontal bending–vertical placement of reinforcement is needed to resist vertical bending and horizontal placement is needed to resist horizontal bending. All of these types of walls are commonly used, but placement of reinforcing steel, if any, is not always consistent with the direction of bending. The report mentioned earlier on building performance, prepared by FEMA, states that the main cause of failure of masonry walls is the lack of vertical reinforcement.

Low-rise residential and commercial buildings often combine masonry walls with wood floor and roof systems and use the floors and/or roof to both stabilize the masonry wall and transfer the forces through diaphragm action to other elements. These walls normally are not designed as cantilevers; they rely upon the floor or roof diaphragm to stabilize the masonry wall. If the upper horizontal support is lost, the masonry wall will topple. When the masonry wall extends only to the ceiling, as is common for gable-roof buildings, the gable-end plate develops bending due to surface pressures in the gable portion, as well as pressures on the masonry portion. Thus stability of the wall is dependent upon this plate. Here, again, the gable-end plate must be continuous between points of

Table 11 WIND RESISTANCE CONSIDERATIONS FOR MAIN-FRAME DESIGN

Rigid or Braced Frames	• Configuration for complete load path, with lateral loading in both directions
	• Adequate strength for both wind and gravity loads, and loading combinations
	• Connections to diaphragms and/or collectors in all directions
	• Anchorage to foundations to resist sliding, uplift and overturning
	• Dead load stabalizing moment minimum of 1.5 times wind overturning moment
Shear Walls	• Number, orientation and location of walls to provide complete stability and minimum torsion for building
	• Connections to diaphragms and/or collectors in all horizontal directions
	• Shear capacity adequate for all loads
	• Details for transferring forces through floor diaphragm to shear wall(s) below
	• Dead load stabalizing moment minimum of 1.5 times wind overturning moment
	• Anchorage to foundations to prevent sliding and uplift
	• Foundation heavy enough to stabalize wall
	• Bearing shear walls work better than non-bearing shear walls
Diaphragms	• Avoid long, slender proportions in plan between vertical elements
	• Monolithic deck is better than one composed of pieces
	• For deck composed of pieces, provide adequate attachment to prevent sliding between picees in order to develop shear resistance and prevent excessive distortion
	• Continuous, unborken chords around full perimeter of diaphragm
	• Connections to vertical elements for all horizontal directions
	• Details for transferring forces through floor diaphragm to shear wall(s) below
Collectors	• Attachment to wind surface material for pressures both inward and outward
	• Positioned to meet strength limitations of surface meaterial
	• Continuous from end to end
	• Adequate strength in bending, tension and compression
	• Connections to main-frame to resist forces in all horizontal directions

its lateral support. The lack of plate capacity and plate movement resulting from the displacement of gable-end trusses were causes of major structural damage to some buildings in the hurricanes mentioned. This sort of structural damage is clearly due to progressive failure, and the extent of damage depends upon the extent of loss of supporting members.

One advantage of masonry is its resistance to debris damage. Debris penetration through masonry is relatively infrequent, although superficial damage to coatings, such as stucco, may occur.

Light-gauge Metal Systems. Hurricanes Andrew and Iniki also exposed weaknesses in light-gauge metal systems and pre-engineered metal buildings. Resulting damage, again, was progressive, with total building collapse in a few instances.

Observations of failed light-gauge metal roof decks reveal that tearing of metal at locations of button welds (a common method for securing the deck to roof trusses) is typical. It is unclear whether this tearing is the initial mode of failure or follows localized buckling of the deck. Regardless of the sequence, lateral stability of supporting roof trusses or other light-gauge metal beams may be lost when the deck is blown away. Here again, partial or total roof collapse may follow. Mechanical connectors, such as self-tapping screws with washers, provide improved performance.

Corrections for the aforementioned failure modes mainly require increased strength or provision of redundant bracing members. Localized deck buckling can be prevented by increasing the deck thickness or increasing the metal thickness (lower gauge number). The second option is usually best, because deck

Table 12 WIND RESISTANCE CONSIDERATIONS FOR DESIGN OF COMPONENTS AND CLADDING

Roof Surfaces	• Sufficient roofing weight or mechanical attachment to deck to resist uplift
	• Design for higher pressures near edges
	• Roofing backing (deck) with adequate strength in bending for pressures upward or downward
	• Resist additional uplift forces at roof overhangs
	• Deck connections to supporting elements (joists or purlins) for forces upward or downward
	• Supporting elements designed for combined wind and gravity loadings
Wall Surfaces	• Cladding or panels attached to resist lateral forces inward and outward
	• Design for higher pressures near corners
	• Surface backing continuous and of adequate strength to resist forces inward or outward
	• Backing connections to collectors or main-frame for forces in all horizontal directions
	• Surface adequate to resist missile impact
Windows and Doors	• Glass surface area proportioned for strength or subdivided with mullions
	• Glass shatter-resistant or protected with shutters
	• Glass frame of adequate strength
	• Window and door frame connections to collectors or main-frame for resistance to forces in all horizontal directions
Collectors	• Attachment to wind surface material to resist pressures inward and outward
	• Positioned to meet strength limitations of surface material
	• Adequate strength in bending, tension and compression
	• Continuous from end to end
	• Connections to main-frame to resist forces in all horizontal directions

depths available sometimes are limited. Closer spacing of button welds will also provide increased resistance to the uplift. Provision of lateral support for roof trusses or beams is most easily accomplished by use of bridging or other bracing rather than relying solely upon the metal deck for this function. It should be noted that uplift pressures may cause reversed stresses in roof beams and trusses, which means that compression stresses are possible in both flanges (chords) of these members. Both edges of the members therefore must be supported laterally.

PRACTICAL CONSIDERATIONS FOR WIND-RESISTANT BUILDINGS SUMMARIZED

In order to provide useful information regarding wind-resistant construction, an outline of considerations is provided in table form with this monograph. Not all of these considerations will apply to every building. The items of consideration have been drawn from the authors' experiences in dealing with architectural applications for lateral force resistance in buildings.

The list of considerations has two major divisions–considerations dealing with overall building behavior, that is, the main-frame–and considerations dealing with assemblies and connections that have demonstrated vulnerabilities to wind damage. Sub-categories further separate the divisions by pertinent topics.

The list of considerations places one significant burden upon any user. The information is of value only if you fully identify those elements of your project that participate in resisting wind forces.

Aspect ratio
A proportion of building height to least plan dimension.

Basic wind speed
Wind velocity at a height of 33 feet in open terrain conditions, and normally obtained from a wind map.

Bernoulli's Equation
A mathematical expression relating velocity and pressure which provides the basis for defining wind pressure.

Boundary layer
Pertaining to wind, that region from the ground surface upward in which effects of terrain roughness (friction) impedes wind velocity. Height of the region above the ground increases with increasing roughness of the terrain.

Boundary-layer effect
Reduction in velocity of the wind, ranging from zero at the ground surface to unimpeded velocity (full velocity) at gradient height.

Collector
A structural element or component which receives surface pressures and transfers the resulting forces to the main-frame, either directly or through other building elements.

Components and cladding
The particular term used to define building surfaces where wind pressures exceed average pressures, and for which additional design pressure must be determined.

Components
Building elements which function as load-carrying elements and which may carry main-frame forces or only localized surface forces, or both.

Dampers
A general term used to describe mechanical devices attached to the structural system and which are for the purpose of reducing building lateral motion resulting from lateral loading.

Design velocity pressure
That equivalent static wind pressure determined for a particular building, a particular height, a particular building use, and a particular design purpose–i.e., main-frame or components/cladding–and which includes pressure coefficient applicable to the surface and gusting effects.

Diaphragm
A building structural element, usually positioned horizontally such as floor or roof system, which transfers lateral loads (forces) to vertical elements of the main-frame.

Eaves beam
A term often used to describe a horizontal (or nearly horizontal) collector element positioned at the top of a wall (or top of a gable-end wall) and which collects lateral forces for transfer to vertical elements.

Exposure category
A classification of terrain roughness conditions typically used in codes and standards for establishing variations in wind velocity for different building sites and which reflects boundary-layer effects.

External pressure
Wind pressure which is developed on external surfaces of a building.

Fastest-mile wind speed map
Wind speed map based upon fastest-mile wind records.

Fluctuation
Variation of wind velocity over time -- seconds, minutes, hours, days, months, seasons, or years. Building response is concerned with fluctuations measured in seconds.

Gradient height
That height above the ground surface above which the boundary layer effect disappears.

Gradient velocity
Wind velocity at gradient height -- i.e., wind at full velocity.

Gust Response Factor
A coefficient which accounts for effects of gusting and used in standard procedures of codes and standards.

Glossary of Terms

Hertz
The frequency of vibration (cycles per second) pertaining to the dynamic response of a building.

Horizontal element
An element of the primary lateral force resisting structural system which is positioned horizontally (or nearly so) and which collects forces for transfer to vertical elements.

Importance factor
A coefficient reflecting the importance of building use or occupancy which is a multiplier in the velocity pressure equation and which effectively may increase or decrease the design velocity pressure.

Internal pressure
Wind pressure which is developed on internal surfaces of a building.

Inward pressure
A term often used to describe positive pressure on a surface. This is not the same as pressure acting inward with respect to the building, although they often are the same.

Leeward surface
A building surface on the downwind side which is positioned perpendicular (or nearly so) to the direction of wind flow.

Main-frame
The primary lateral load resisting structural system of a building which includes both horizontal and vertical elements.

Mean recurrence interval
A time span (typically in years) during which wind velocity is not expected to exceed a specified velocity.

Negative pressure
Wind pressure acting away from and normal to a surface.

Outward pressure
A term often used to describe negative pressure on a surface. This is not the same as pressure acting outward with respect to the building, although they often are the same.

Overturning moment
The tendency for rotating a building (or element) about a base point as a consequence of a lateral force. Net overturning moment is the algebraic sum of the overturning moment and the stabilizing moment and, for the final wind design, must be less than zero.

Positive pressure
Wind pressure acting toward and normal to a surface.

Power law
An equation which relates wind velocity and height based upon an exponential relationship.

Pressure
Force per unit area (usually psf for buildings) acting on a surface.

Pressure coefficient
A coefficient which, when multiplied by a velocity pressure, approximates the average pressure on a surface. Coefficients vary with surface orientation and location relative to direction of wind flow and size of the building or surface.

Pressure vector
The magnitude and direction of a pressure–i.e., acting toward or away from a surface.

Primary wind-load resisting system
Another name for the main-frame.

Probability of recurrence
A statistical estimate, based upon mathematical probability, of exceeding a specified wind velocity during a specified time period (usually measured in years).

Progressive failure
A description of building failure which occurs incrementally as a consequence of failure of elements of increasing importance in the structural system.

Recurrence interval
That time period which elapses between occurrences of the same wind velocity.

Relative stiffness
A value which relates the stiffness of one structural element to another based upon deformations.

Roof edges
As related to standard wind analysis procedures, a unique portion of a roof surface where localized increased pressures are considered.

Shear wall
A vertically positioned element of a main-frame which transfers lateral forces primarily by shearing action from diaphragm or collector beams to the foundation of a building. A shear wall resists lateral forces only in a direction parallel with the wall.

Side surface
A building surface which is positioned parallel (or nearly so) with the direction of wind flow.

Stabilizing moment
That rotational effect about a base produced by permanent gravity loads which acts counter to the overturning moment.

Stagnation pressure
A theoretical wind pressure produced on the windward face of an obstruction of infinite expanse so as to allow no flow past the obstruction.

Stepped-pressure profile
A wind pressure profile, used in practice, based upon constant pressure values at selected increments of height and which provides a simplification of an exponential profile.

Terrain roughness
A term used to describe different degrees or types of ground surface conditions which affect wind flow in the boundary layer.

Three-second gust wind-speed map
Wind speed map based upon three-second gust velocity wind records.

Turbulence
Mixed (garbled) air flow, such as eddy currents, in near proximity to surfaces where smooth air stream flow has been altered due to the surface(s). This mixed-up air flow may produce significant pressure reversals and consequent lateral deflection of tall (wind-sensitive) structures.

Velocity
Wind speed, usually in miles per hour.

Velocity pressure
A particular pressure used in wind analysis (in some references called the stagnation pressure or basic design pressure) which may include considerations of height, building importance, and local terrain variations. This is the pressure used to establish a particular design pressure and so is dependent upon height for the surface being considered.

Velocity profile
A graph that plots wind velocity as a function of height.

Vertical element
An element of the primary lateral force resisting structural system which is positioned vertically and which receives forces from horizontal elements or components.

Windward surface
A building surface which faces toward the oncoming wind flow and is positioned perpendicular (or nearly so) to the direction of wind flow.

BIBLIOGRAPHY

ASCE Standard: *Minimum Design Loads for Buildings and Other Structures*
New York, American Society of Civil Engineers, 1993

ASCE Standard: *Minimum Design Loads for Buildings and Other Structures*
New York, American Society of Civil Engineers, 1995

Building Performance: Hurricane Andrew in Florida
Washington, DC, Federal Emergency Management Agency, 1992

Building Performance: Hurricane Iniki in Hawaii
Washington, DC, Federal Emergency Management Agency, 1993

Crawley, S.W., and R. Dillon. *Steel Buildings: Design and Analysis*
New York, John Wiley, 4th Edition, 1993

Crawley, S.W., and D.B. Ward. *Seismic and Wind Loads in Architectural Design*
Washington, DC, American Institute of Architects, 1990

Keith, E.L., and J.D. Rose. *Hurricane Andrew: Structural Performance of Buildings in Southern Florida*
Tacoma, WA, American Plywood Association, 1992

National Building Code
Chicago, Building Officials & Code Administrators International, Inc. (BOCA), 1993

Parker, H., and J. Ambrose. *Simplified Engineering for Architects and Builders*
New York, John Wiley, 8th Edition, 1993

Standard Building Code
Birmingham, AL, Southern Building Code Congress International (SBCCI), 1993

Uniform Building Code
Whittier, CA, International Conference of Building Officials (ICBO), 1994

NCARB PROFESSIONAL DEVELOPMENT PROGRAM WIND FORCES QUIZ ORDER FORM

If you received this monograph without a quiz, or if you would like to order additional copies of the quiz, please complete this order form and return it to NCARB. When you pass the quiz, NCARB will issue you a certificate of completion for 14 contact hours of continuing education credit. If you are an AIA member, NCARB will report your passing score to AIA. You will receive 42 AIA Learning Units for this monograph and quiz.

SHIP TO:

Name ____
Please print

Firm ____

Address ____

____

City ____

State/Province ____________________ Country ____________________

ZIP/Postal Code ____

Daytime Phone ( ) ____

FEES

- ☐ Wind Forces Quiz .... $70* x quantity_____ = $__________
- ☐ Please add U.S. $20 for international shipments (excluding Canada)

* Fees are current as of July 1998, and are subject to change.

PAYMENT

☐ Check enclosed in the amount of $ __________ *All checks must be drawn on a U.S. bank.*

☐ Charge my ☐ Visa ☐ Mastercard ☐ American Express

____
Account number

____
Expiration date

____
Signature

Mail order form and payment to
Professional Development Program, National Council of Architectural Registration Boards,
1735 New York Ave., NW, Suite 700 Washington, DC 20006

WF